The Author

Johnson Stanley is working in ICAR as Scientist at Vivekananda Institute of Hill Agriculture, Almora, Uttarakhand. A student of TNAU, Coimbatore, he obtained many medals/ awards and scholarships both in M.Sc. and Ph.D. As a scientist, he is working in various projects and published 30 research papers and book chapters. Wildlife always being a fascination from childhood got enhanced by working in hill station experiencing severe crop damages by them.

Animal and Bird Pest Management in Agricultural Land

Animal and Bird Pest Management in Agricultural Land

Author

Johnson Stanley

2015

Daya Publishing House®

A Division of

Astral International Pvt. Ltd.

New Delhi - 110 002

Cataloging in Publication Data--DK
  Courtesy: D.K. Agencies (P) Ltd. <docinfo@dkagencies.com>

  Stanley, Johnson, author.
Animal and bird pest management in agricultural land / author, Johnson Stanley.
    pages cm
    Includes bibliographical references (pages   ) and index.

    ISBN 9789351306764 (International Edition)

    1. Vertebrate pests--Control--India. 2. Agricultural pests--Control--India. I. Title.

DDC 632.660954   23

Published by   :   **Daya Publishing House®**
                              A Division of
                              Astral International Pvt. Ltd.
                              – ISO 9001:2008 Certified Company –
                              4760-61/23, Ansari Road, Darya Ganj
                              New Delhi-110 002
                              Ph. 011-43549197, 23278134
                              E-mail: info@astralint.com
                              Website: www.astralint.com

Lesser Setting   :   **GRB 7color Service**

Printed at   :   **Thomson Press India Limited**

PRINTED IN INDIA

Preface

Agricultural crop damage by vertebrate pests like deer, nilgai, blackbuck, wild pigs, rabbits, elephants and monkeys apart from their attack on human beings have been widely noticed in India. Though heavy crop damages realized, management aspects are not properly studied or reported from India. The aim of this book is to review the management aspects available for vertebrate pest management around the world and to facilitate the pest management personals to have the information readily available for selection and deployment according to their needs. The idea of writing a book originated from a meeting on Hill Agriculture headed by the Vice-Chancellor, G.B. Pant University of Agriculture and Technology, Pantnagar, Dr. B.S. Bisht and the Director, Vivekananda Parvatiya Krishi Anusandhan Sansthan (Indian Council of Agricultural Research), Almora, Dr. J.C. Bhatt. The severity of vertebrate pest damage to agricultural crops and the need for their management was felt by all and I have been given with the task of reviewing the literature available on different aspects of vertebrate pest management and come up with a book, which may be the base for further studies in this aspect. The work was accomplished by the divine grace of God Almighty. The help provided to me by my friends and relatives especially my wife, Dr. G. Preetha are thankfully acknowledged. I hope the compilation made, will help the personals working in this aspect to frame effective management methodologies which pave a way for successful deployment for efficient vertebrate pest management.

This book titled *Animal and Bird Pest Management in Agricultural Land* reviews the different methods for the repulsion and management of vertebrate pests and is broadly divided into two sections. The first section explains the methods by which these pests are waded off using repulsive or scaring techniques and the chapters within this section are based on sensation and perception of the pest to these repellers. The second section is on the management of vertebrate pests and further classified in to chapters as mechanical, electrical, chemical and biological management with resistant or tolerant crops/varieties. For an effective vertebrate pest management four 'knows' are to be known. 1. know the pest, 2. know the damage potential, 3. know the management and 4. know the cost economics.

Scarers can be effectively used to wade away vertebrate pests because they are highly mobile in nature and flea from areas of danger. But a combination of different scarers (visual, auditory, olfactory, tactile and gustatory) may be used for an effective management. In case of pest control, poisoning can be a very easy and economic way of killing and eradicating the pests, but should include environmental costs (chapter 8.4). Exclusion can be a good, cost effective and environmentally safer technique (chapter 6.4). Trapping and relocation either using mechanical traps (chapter 6.5) or using tranquilizers (chapter 8.3) may be carried out for animals with greater values. Preventing birth rather than allowing it to born, grow and establish and then managing can be a better option (chapter 8.2). Releasing predators, 'cat for rat', can be a practice wherever feasible (chapter 9.5). Resistant varieties may not be available for all sorts of pest problems but can be a viable option with a long term approach (chapter 10). With all these background, one cannot directly go into the capture or kill of a pest animal because of strict legislative laws on wild life. So, a chapter on regulatory measures and wildlife act is also included (chapter 11). The last and the important chapter deals with the integrated vertebrate pest management (chapter 12) with the combination of available and effective techniques aiming damage reduction rather than pest control.

- Johnson Stanley

ICAR- VIVEKANANDA PARVATIYA KRISHI ANUSANDHAN SANSTHAN
ALMORA – 263 601, UTTARAKHAND

Dr. J.C. BHATT

Director

Foreword

Crop damage by vertebrate pests (mammals and birds) is widespread in our country. Though the crop damage caused by these pests account for a huge loss, management aspects of vertebrate pest has not been given proper attention as that of insect pest management. The crop damages by vertebrate pests are neither properly documented nor studied or reported elsewhere in our country. This book comes therefore at an opportune time to enlighten people about the different vertebrate pest management techniques which can be adopted to minimize damages by them.

Vertebrate pest management, in-fact have a multidimensional approach. A noxious pest for one group is some sort of a recreational species with aesthetic value for some other. People who have strong animal welfare concern want better management techniques with reduced suffering to the animal. Economists argue spending on vertebrate pest management should give economic returns and farmers like to sure that the expenditure incurred for pest management is not wasted by rapid reinvasions from nearby areas. Thus the author had a difficult task in considering all these contemporary views while preparing this manuscript. I found the book is carefully written suggesting an integrated vertebrate pest management emphasizing on managing the pest damage rather than simply reducing the pest population.

The book is crammed with useful tips on vertebrate pest management. The views were properly supported by taking details from scientific literature published, formatted in two chapters as scarers and repellers and the second one on management techniques. Of all the methods, a long term habitat modification or management can be the best solution for vertebrate pest problems. "Don't destroy their habitat they will never destroy your farm" can be the mantra for vertebrate pest management.

Overall the management practices suggested is to reduce the vertebrate pests below some economic levels and holding it there. No doubt, the book is a knowledge tool to management personals and scientists working on these aspects and also to the farmers and land managers to reduce the agricultural losses by them. I congratulate the author for coming out with novel ideas in the form of a book.

27.10.2014

(J. C. Bhatt)

Ph: 91-5962- 230208 (O); 230130 (R); 231539 (F)

E-mail Office: vpkas@nic.in

Personal: jagbhatt@yahoo.com

Contents

List of Figures

Introduction

The old adage about sowing wheat,

"One for the rook and one for the crow, one to rot and one to grow"

is of very ancient origin, held true for centuries and shows the importance of vertebrate pest and the damage caused by them. It is true that the bird damage starts before the plant grows in the field and that too in some cases, double the portion of the plant that emerges from seed will be taken away by pest birds. Birds damage fresh sown (digging out seed/ seedling) to ripened crop of maize, lentil, pea, barley and wheat (Brooks and Ahmed, 1990). Often, damages are more severe to cropping fields near to their roosting sites. Some birds also damage soft fruits such as guava, orange, mango and loquat (Hafeez *et al.*, 2008).

In case of mammals the situation is still worse that they allow the plant to grow and eat away the produce of economic importance and the damage is considerable. Raiding by large mammals includes trampling, uprooting, eating whole plants or their economic parts apart from direct human conflicts are well documented. Smaller animals such as monkeys, baboons, porcupines and bush pigs also raid crops by removing above ground (maize cobs) and below ground (potatoes, ground nuts etc.) economic parts. Agricultural crop damage by deer, nilgai, blackbuck, goats and wild pigs have been widely reported (Schultz, 1986). Crop depredation by elephants is very heavy especially in areas near forest covers and mountains. Langur and monkey occur commonly in forests interface agricultural fields, orchards, villages, townships and cities. Both the species have adapted to human habitation and depend on agricultural and horticultural crops and food handouts by local people. Sometimes their attacks on human beings are also noticed owing to their aggression or adaptive behaviour. In addition to damaging habitats, feral pigs, buffaloes, horses, cats, dogs and rats are also prominent reservoirs for exotic and endemic diseases and parasites that can affect native wildlife, stock and humans (Dickman, 1996; Levett *et al.*, 1998; Thumen *et al.*, 2002; Watts, 2002). Thus there is a need to control/ manage these vertebrate pests to reduce crop damage and increase crop production especially in vulnerable areas *viz.*, fields near woody forests and mountain terrains.

I. Scarers and Repellents

Mostly vertebrate pests are highly mobile and thus can be easily frightened off or wade off from the area to be protected using scarers and repellents. Novel substances in the field (neophobia), unnatural sounds (fear), predator like odour (instills fear in herbivores), irritants (causes pain, making area less hospitable), distasteful substance (changes the palatability of food items), food aversive substance (causes illness after ingestion and thus rejected) are denoted as repellents. Repellent substances and devices cause pest species to avoid otherwise attractive or palatable to them (Rogers, 1978; Mason *et al.*, 1989). On the basis of animal sensation and perception, repellents can be of visual (flags, balloons, lights, scare crow), auditory (crackers, propane guns, biosonics), olfactory (smoke, aerosols), tactile (clay coated seeds) and gustatory (ash dusting, non preferred plants).

Frightening methods and devices/stimuli to prevent animal damage are useful for some pest species and are most effectively used where crops need to be protected for a relatively short period such as just prior to harvest. Further, it is well-documented that the fear response in most animals includes a series of visual, acoustic and possibly olfactory stimuli. So a combination of repellents has to be used to have effective management strategy.

II. Pest Management

On the basis of mechanism of action, vertebrate pest management may be of mechanical which in-turn as scaring (spines), trapping (various traps) and protection (netting), electrical (electrical fence), biological (guard dogs) and chemical which in turn as repellents, immobilizers, sterilants, toxicants etc.

Resistant crops/ varieties: Planting unattractive crops and growing resistant/ tolerant varieties for bird/ mammal pest damage is also an option. This forms the basis for all the pest management approaches, but breeding a crop for vertebrate pest resistance without compromising yield and quality is a huge task to be made with. Use of unfavourable crop or varieties in the borders is a tactic to repel the pest animals.

Integrated vertebrate pest management (IVPM): Vertebrate pest management can be aimed by using a combination of repellent and management practices to be more effective. Surveying for pest damage potential, setting of threshold levels and use of effective techniques are to be practiced in integrated vertebrate pest management programmes. Integrated vertebrate pest management aims at reducing the damage rather than reducing the population of the pest per se.

Munias feeding on rice

Parakeet damage in wheat

Feral pigeons feeding in lentil field

Bird damage to radish pods

Parrot damage in corn

Bird damage in guava fruit

Photo 1. Bird damage in various agricultural crops

Deer damage to cauliflower

Wild pig damage to corn

Porcupine damage in pigeon pea

Rabbit damage in soybean

Wild pig damage in upland rice seedling

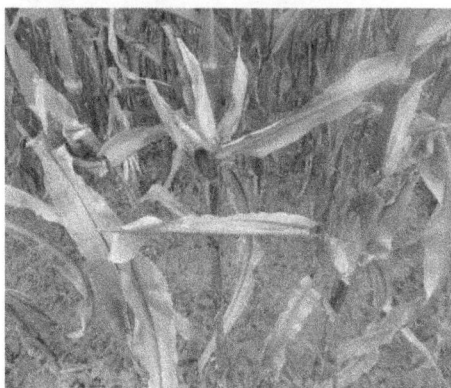

Monkey damage in corn

Photo 2. Animal damage in various agricultural crops

Section I
Scarers and Repellers

Chapter 1

Visual Scarers

Vision-based deterrents present a visual stimulus that is novel, startling or that the animal associate with danger. The danger can be a predator, a simulated predator and the results of a predator attack (dead bird or model thereof). These are often inexpensive and effective, if only for short periods. Functionally visual scarers cause startle responses, as do aposematic colours (red, orange, silver: Reidinger and Mason, 1983) and cues associated with the predators (eyespots, raptor models, hawks: Inglis, 1980, Conover, 1982, Inglis *et al.*, 1983). Animals and birds are frightened by fire. Fire and fire imitators as that of flash and laser lights can make the animals and birds to avoid the place. Hanging of dead birds is a common practice to scare birds. However these startle responses eventually diminishes (often within days or a few weeks) as a function of several variables, including weather conditions, animal population and availability of nearby unprotected foods (Feare *et al.*, 1986).

1.1. Flags and Streamers

Flags and streamers are caught in the wind and the resulting movement startles birds and animals, often causing them to flee the site. Flags hung from poles or wires have occasionally been used to frighten birds from agricultural fields (Marsh *et al.*, 1991). Birds may be initially repelled, especially if the flags flutter in the wind, but in due time the birds become habituated and even use the poles and wires as perches (DeHaven, 1971). Brightly coloured eyes are usually painted in flags to enhance their efficiency. Their effectiveness has usually been limited especially when used alone.

A best of protective method, namely 'Continuous string flagging', is used in many places against horned larks, *Eremophila alpestris*. It is effective for larks because unlike other birds which start its depredation in the borders or near fences, horned lark activity is well noted in the center of the field (Koehler, 1962). Neff (1948) described two methods of flagging agricultural fields to protect crops from damage by horned larks. One technique involved attaching strips of cloth or paper to the top of stakes spaced at intervals of 20 to 25 ft. Another in which, twine was stretched across a field in parallel rows 20 to 30 ft apart and attached to 4 ft tall poles and small paper or cloth streamers were tied to the twine at 5 ft intervals. In rye and winter wheat fields, simple white flags are reported to significantly reduce snow geese, *Chen caerulescens* damage, revealing this as an economic and efficient method of geese management (Mason *et al.*, 1993). Mason (1995) reported white plastic flags repelled snow geese from damaging salt marshes, *Spartina* sp. as revealed by significantly fewer feces in treated areas with respect to control. Flagging affords crop protection in direct ratio to the density of stakes and flags. For complete protection the stakes cannot safely be placed more than 25 ft apart and in case of persistent attacks should not be more than 20 ft apart (Koehler, 1962).

Coloured flags were also tried as a means of reducing damage by red-billed quelea, *Quelea quelea* feeding in a rice fields (Manikowski and Billiet, 1984). Flags of 20x20 inch cloth pieces were attached to 6.5 ft tall poles. Fewer quelea were reported to be observed in flagged plots than in adjacent untreated plots and white and red flags as more effective than black, yellow or blue flags. In another study, black seems to be the most effective colour and flags of 60x90 cm size can easily be seen and may move well in the wind (Bishop *et al.*, 2003). Mason and Clark (1994) demonstrated that both black and white flags placed at a density of 2.47 flags/ha reduced snow goose use of fields, based on counts of goose faeces, but did not deter geese completely. White flags placed at the same density in crop fields reduced goose numbers and even deterred geese from fields where they had been grazing for the previous few weeks (Mason *et al.*, 1993). However, McKay and Parrott (2002) found that white flags placed at 25 m intervals in a regular grid were ineffective in reducing mute swan grazing on oilseed rape. Silver and red Mylar streamers can be used as an alternative to flags and were tested by Mason and Clark (1994). They were found to delay the onset of grazing by snow geese, compared to fields with no streamers or flags, but proved less effective in reducing goose use than either black or white flags. Mylar flags have also been used to deter Canada, brent and snow geese from agricultural fields (Heinrich and Craven. 1990; Summers and Hillman, 1990; Mason and Clark, 1994). Mylar flags (15 cm x 1.0 m) attached to wire or lathe supports positioned at 6 m intervals at nesting colonies did not repel herring gulls, *Larus argentatus* and ring-billed gulls, *L. delawarensis* whereas it found to repel the same birds from a loafing site. When the loafing area is small and near to frequently visiting pond, then mylar flags are found not effective (Belant and Ickes, 1997).

Rags tied between tall maize plants effectively protected the maize crop from red-winged blackbird damage up to 30 days, after which there were signs of habituation (Cardinell and Hayne, 1945). A device consist of two (0.9x0.6 m) flags attached to a horizontal 1.9 m boom which when mounted to a vertical post or stake will rotate in the wind was used to wade off pigeons. The flags are made from yellow nylon with an enormous (0.7x0.5 m) red eye and a menacing (0.2 m) black pupil on both sides. This device is reported to be temporarily effective in repelling pigeons (Woronecki, 1988).

Two mylar flags (15 cm x 1 m) attached to lathe positioned at each of 5 sites did not reduce the number of deer intrusions into feeding stations or the amount of corn consumed relative to feeding stations without mylar flags. Thus it was concluded that mylar flags are ineffective for deterring white-tailed deer, *Odocoileus virginianus* from feeding areas during winter especially at high deer densities of more than 21/km^2 (Belant *et al.*, 1997a). In Zambia, white squares of cloth have been hung from string tied around fields of crops with the hope that a fluttering object might alarm approaching elephants (Osborn, 2002). Wire fences with white, flying, flashing ribbons or plastic strips that produce scaring sounds can be used in and around crop fields to deter wild pigs (Chauhan *et al.*, 2009).

In summary, flags and rags are a useful scaring technique, being cheap and easy to deploy. However, their success depends on alternative feeding sites being available nearby. Although not totally effective in completely eliminating bird pests, reducing numbers by dispersing birds over an area can reduce localized crop damage.

1.2. Balloons, Kites and Raptor Models

Balloons tethered in a crop are an inexpensive method of bird deterrence. Various balloons and balloon-like objects have been tested for use against birds in protecting crops or dispersing roosts and most have achieved some success (Hothem and DeHaven, 1982; Mott, 1985). Experiments involving young chickens and adult European starlings, *Sturnus vulgaris* demonstrated the avoidance reactions to eyespot patterns under laboratory conditions (Scaife, 1976; Inglis *et al.*, 1983). Helium filled vinyl balloons with five eye spots was reported as a bird deterrent reducing grey starlings, *Sturnus cineraceus* damage in grapes, cherries and peaches (Shirota *et al.*, 1983). Toy balloons tied to branches of the trees deterred starlings from causing damage to cherries and blueberries but robins and Baltimore orioles were seen to continue feeding only a few feet away (Bishop *et al.*, 2003). Balloons are also used in combination with Mylar reflecting tape to enhance its efficiency (Bruggers *et al.*, 1986).

In a study to deter boat-tailed grackle, *Quiscalus major* from orange orchards, Avery *et al.* (1988) reported that the presence of the eyespot balloons have increased the mean distance of the birds from the trees and reduced the frequency of tree area used. Some birds are found to land on the nearby areas but not found to peck the oranges but fly away quickly. A white balloon with red and black eyespots was reported to keep birds at a greater distance from the orange trees whereas a black balloon with orange and yellow eyespots did not repel the birds from the grove (Avery *et al.*, 1988). Not only balloons, beach balls painted with eyes are also reported to significantly reduce the house sparrows, *Passer domesticus* from visiting a bird-feeding table (McLennan *et al.*, 1995). He further evaluated eyespot beach balls and commercial bird repellent ball as a bird deterrent in vineyards and found commercial balls are much effective in 10, 20, 30 and 40 m distances. The deterrent effect of both balls decreased with distance from the table and was negligible at ~40 m. Deterrence was found to increase 10% by illuminating the eyes with a rotating halogen light. In the first three weeks, the balls repelled 90% of yellow-headed blackbirds (*Turdus merula*) house sparrow, (*P. domesticus*) and starlings except song thrushes (*Turdus philomelos*) which had started to ignore them in the second week.

Eye-spots on balls might scare birds because they mimic the eyes of a large raptor or mimic the frontal threat display of an aggressive conspecific (Inglis, 1980). Various experiments show that eye-spots are effective only if they are large and are arranged in pairs so that they appear to stare, have a three-dimensional appearance and have a strong contrast between pupil and iris (Scaife, 1976; Inglis *et al.*, 1983). Indeed, once habituation has occurred, eye-spot balls could even attract birds by signaling the whereabouts of good feeding areas (Inglis, 1980). Monitoring of bird damage to natural bunches of Riesling grapes within vineyard blocks showed that the rate of starling damage to the crop was reduced significantly by the eye-spot balloon relative to Peaceful Pyramid®, a commercial visual bird scarer and control plots. Nevertheless, after 24 days of monitoring, birds in the balloon plot and pyramid plot had damaged 75 and 84% of the grapes, respectively. In another trial, clusters of table grapes were attached to vineyard wire and found both the balloon and pyramid reduced bird damage to clusters only within 15 m of the device, but had no measurable effect on clusters further away thus rendering it uneconomical (Fukuda *et al.*, 2008). Horikawa *et al.* (1988) examined the response of rufous turtle dove, *Streptopelia orientalis* to several kinds of eyespot patterns drawn on a whiteboard in a flight cage but did not confirm any difference in bird behaviour. Balloons themselves were found to have a scaring effect on turtle doves with or without eyespots when alternate source is available, thus revealing the scaring effectiveness of eyespot patterns as insignificant. Among eyespots, avoidance tendency was noticed more for paired red eyespot balloons than yellow (Nakamura *et al.*, 1995). Reduction in damage due to *S. orientalis* was observed in soybean, when several kinds of eyespot balloons were placed in open field (Yui, 1988) but no deterrence was observed among feral pigeons, *Columba livia* (Shimizu *et al.*, 1988).

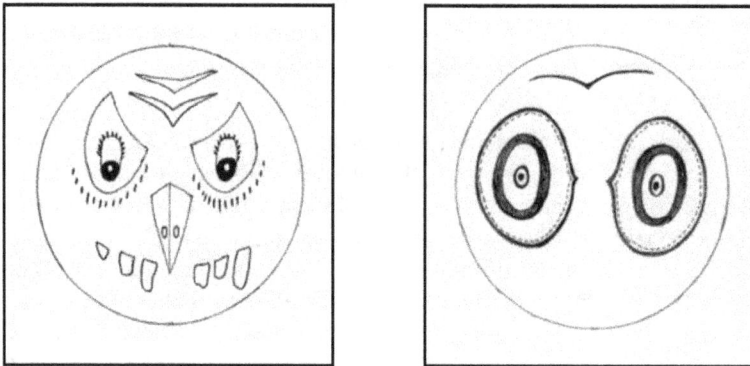

Figure 1.1: Eyespot balloons

Although balloons are reported to be effective, it is not known that how many balloons are needed to protect a particular area. In one field test, a helium filled balloon of 2.6 m in diameter floated about 15 m above ground level was reported to successfully reduce bird depredations over approximately 3.5 ha (Shirota *et al.*, 1983) whereas Fukuda *et al.* (2008) reported that balloons are effective for 15 m distance only. The specific details (balloon size, style, number, etc.) relating to application of this technique may vary from place to place and bird to bird. Most of the reports showed eyespot balloons as effective for 4 to 7 days, since birds quickly habituate to them. These balloons with combination of other scaring techniques may be effective against bird damages.

Kites

Kites are largely reported to be ineffective at deterring birds (Bishop *et al.*, 2003) but worth testing. Hothem and DeHaven (1982) tested raptor mimicking kites suspended from helium balloons in a vineyard has contributed for only a slight decrease in a small area, but the damage elsewhere in the vineyard increased. But woodpigeon, *Columba palumbus* damage to spring cabbage was greatly reduced when a large kite was flown over the fields. Protection continued for over 3 months with no sign of habituation to the kite by pigeons. Woodpigeons avoided flying or settling within 250 m of the kite. Damage in fields with a gas banger exceeded that in fields with a kite, especially in severe winter weather. It was further reported that the kites should be launched early in the morning before the arrival of birds for effective management of woodpigeons (Haque and Broom, 1985).

Raptor models

One approach to bird damage control has been to frighten birds from an area by exploiting their innate fear of raptors. To this end, gas balloons tethered live raptors (Davids, 1960) and raptor models (Brown, 1974; Blokpoel, 1976) have all been tried. Like any new object placed in the environment, they may be avoided by other birds for a few hours or days. However, the pest species soon learns that the models are no threat and pay no attention to them (Marsh *et al.*, 1992). For some bird species the avoidance response to avian predators appears to be, in part, a learned behaviour. Juvenile grey jays, *Perisoreus canadensis* showed little response to a model great horned owl, *Bubo virginianus*, whereas adult jays reacted intensely (Montevecchi and Maccarone, 1987).

Models of owls are often promoted and used unsuccessfully in an attempt to repel pest birds. Although several investigators have found various combinations of raptor-silhouettes and balloons to be ineffective for reducing damage by birds (Rowe, 1971; Messersmith, 1975). Hothem and DeHaven (1982) has given different kinds of effective raptor models. The eagle model, a plastic kite with a 0.76 m wingspan and the likeness of a Peregrine falcon, *Falco peregrinus*, a delta-shaped kite with a 1.65 m wingspan with a wooden-dowel framework, and the blue body and wings and the white head and tail made of polyester and cotton cloth sewn together to resemble an eagle, simple spherical balloon of blue, white, orange, red and yellow shades and an orange tetroon, a tetrahedron-shaped balloon made of polyurethane with a volume of 0.5 m³ and used with some success.

Conover (1979) evaluated the response of birds to raptor models of sharp shinned hawk, *Accipiter striatus* and a goshawk, *A. gentilis* at five artificial feeding stations and a blueberry plot. They initially deterred birds but most habituated to the models after only 5 to 8 hours. Blue jays (*Cyanocitta cristata*) and starlings were deterred more than mockingbirds (*Mimus polyglottos*), mourning doves (*Zenaida macroura*) and house finches (*Carpodacus mexicanus*). Conover (1985) also evaluated a great horned owl model for protecting vegetable crops from crow, *Corvus brachyrhynchos* depredations. Three versions of the model were tested in tomato and cantaloupe plots. The tests were by using an unanimated plastic model, the same model with grasping a crow and the third one is the same as that of second but with battery operated movement in crow's wing. Damage to fruit was assessed during each treatment and compared to damage levels in an untreated plot. The unanimated owl model was reported ineffective. Both animated versions reduced damage by 81% when compared to the control plot.

Although the hawk models significantly reduced the number of feeding birds, they were not as effective as a hawk kite suspended from a helium-filled balloon. Hawk kite reduced the bird predation by 40% for a week in blue berry plots. Although a hawk silhouette suspended from a gas-filled balloon was found to be impractical to use in corn fields (Messersmith, 1975) and ineffective for forest nursery protection (Rowe, 1971), a raptor-mimicking kite suspended using a helium-filled balloon *i.e.* kite-balloon has reportedly been successful in reducing bird damage in grape vineyards to a tune of 70-80% (Hothem and DeHaven, 1982). Woodburne (1979) estimated that a single kite-balloon could protect 2 to 2.5 ha of grape vineyard area. Based on the results of both damage assessments and bird censuses, Hothern and DeHaven (1982) concluded that one kite-balloon per ha deployed for alternate one week periods, reduced losses caused by birds by about 48%. Bird scaring models *viz.,* frightening kites, helical hawk eye balloons and wind powered hawk eyes rotators were combined to visualize their joint effect and installed at equal distances in mango, guava and citrus garden to deter predation of parakeets, *Psittacula krameri* and found less successful than reflecting tapes in reducing the fruit damage (Khan *et al.,* 2011).

Other than human effigies plastic alligator heads, coyote effigies are also used to deter Canada goose (Preusser *et al.,* 2008). Taxidermy mounts of coyotes deployed to move in the wind might be useful as part of an integrated program to disperse Canada geese and other birds from airports (Cleary and Dolbeer, 2005). Models of owls are often promoted in garden catalogs and used unsuccessfully in an attempt to repel pest birds. The use of raptor perches and perching kestrel models on some of the perches was found ineffective in significantly repelling pest birds from vineyards (Howard *et al.,* 1985). Models of snakes apparently are not perceived as a threat as birds show little fear of these. The same is true of other unrealistic models or those without a major biological significance to birds (Marsh *et al.,* 1992). Fake snakes kept in nesting sites of birds are found effective in deterring them from roosting and nesting.

Problems and improving efficiency

The important problem is keeping the balloons aloft. Although kite balloons are apparently more effective during a light wind than during calm conditions (Conover, 1979), winds greater than about 8 km/hr blow tetroons very near the ground. This would probably increase the risk of damage to the kite balloons while reducing its range of effectiveness. The spherical balloons, although more aerodynamically stable in high winds, tend to become inoperable after only a few days. Their longevity might be increased if they were placed in a sheltered location at night or at other times when birds were not a threat and when weather conditions were most severe. If these balloons are used for extended periods, birds become habituated to them and their effectiveness may be reduced. According to Shalter (1978), the relative lack of habituation to recurring predators in nature is, in part, a function of their ever-changing spatial relationships relative to the inanimate environment. He demonstrated that habituation by birds could be reduced by periodically changing the location of the model predator. The models, kites and balloons can be tethered with thick nylon threads with lengths ranging from 16 to 60 m. The kites can be attached each day at a different position on the tether line from 8 to 52 m above the ground and the location of the kite balloon tether point should be moved at least 25 m each day.

Other possible means of improving the effectiveness of kite balloons include: (1) the use of a kite which mimics a bird of prey common within the range of the pest species (Inglis, 1980), (2) the presentation of the kite balloons as infrequently as possible, perhaps only during times of the day when birds are feeding (Slater, 1980), (3) the incorporation of alarm or mobbing calls of the pest species (Inglis, 1980), (4) the reinforcement of threat by supplemental devices such as a gunshot or the presence of a human (Slater, 1980), (5) the use of a live predatory bird to enhance the response to the kites (Inglis, 1980) and (6) the use of alternate bird scarers, using kite balloons for perhaps only 5-7 days followed by conventional devices for a few days before the kite balloons are deployed again (Kimber, 1963).

Conclusion

Hawk and owl models in some circumstances may be more effective than scarecrows, but birds can rapidly habituate to their presence (Conover, 1982). For best results, raptor models should appear lifelike, be highly visible and be moved frequently at the site to help alleviate habituation (Neff, 1979). Since the effectiveness of the kite balloons is apparently species-specific, further study is needed to determine which combination of kites and balloons is most effective against each of the major depredating species of birds. In addition, the range and duration of kite balloon's effectiveness should also be determined for each species under various conditions. Such data should be gathered during whole-season tests on entire farm so that the actual cost-effectiveness can be determined. Eventually, the effectiveness of this technique when used as part of an integrated pest management program should be evaluated.

1.3. Reflective Ribbons

Reflecting ribbons such as Mylar tape (11 mm wide, 0.025 mm thick with metallic red and silver on opposite sides) has been used in attempts to deter birds in a number of circumstances. The tape has a silver metal coating on one side that reflects sunlight, causes a flashing effect and also produces a humming or crackling noise when moved by the wind. Thus, under optimum conditions, reflecting tape produces both unnatural visual and acoustical stimuli for frightening birds (Marsh *et al.*, 1991). In fields, it is usually suspended at parallel intervals above the crop by twisting and stretching it between erect poles.

Reflecting tapes have shown promise in reducing various bird activities in field trials in the United States, Philippines, India and Bangladesh (Bruggers *et al.*, 1986). Tapes as a scaring device act as a combination of visual and exclusion deterrence. They are easy to erect and a wide selection of twines and tapes are readily available. But a varying degree of effectiveness is reported by many authors for different bird species and at different situations. Dolbeer *et al.* (1986a) found that reflecting tape stretched over agricultural crops deterred certain bird species but was ineffective against others; red-winged blackbirds, cowbirds and house sparrows (*P. domesticus*) were generally repelled but goldfinches (*Carduelis tristis*) and mourning doves (*Z. macroura*) showed little reaction. Bruggers *et al.* (1986) reported effectiveness of tape in sunflower and corn against rose-ringed parakeets (*P. krameri*), foxtail millet against munias (*Lonchura* spp.), corn against crows (*Corvus* spp.), sorghum against European tree sparrows (*Passer montanus*) and munias, finger millet and corn against red-winged blackbirds (*Agelaius phoeniceus, Xanthocephalus xanthocephalus*), and sunflower against goldfinches (*C. tristis*). Rice growers in Japan are reported to use reflecting tapes to protect fields from depredations by Java sparrows, *Padda oryzivora* (Bruggers *et al.*, 1986). Summers and Hillman (1990) used red fluorescent tape suspended from poles to protect fields of winter wheat from Brent geese, *Branta bernicla*. They reported reflective ribbons as a cost-effective scaring method for Brent geese when suspended as long lines across the field. However, when all the fields were taped the geese eventually grazed in the taped fields but at a lower grazing intensity than in years when this scaring method was not employed. Reflecting tapes are reported as effective tool against parrots and jungle crows damaging sunflower and ground nut in Pakistan (Hussain, 1990). Reflector ribbons reduced the occurrence of house crows in wheat fields at the seedling stage from 27 to 2. In maize field, it has reduced the number of house crows, parakeets and mynahs by 73.3, 81.8 and 37.5%, respectively (Hafeez *et al.*, 2008). McKay and Parrott (2002) used a combination of hazard warning tape with twine to deter mute swans, *Cygnus olor* from grazing oilseed rape. Although there was evidence that swan grazing was reduced during the first eight weeks, it was not effective in reducing the total amount of grazing over the four month long crop. However, reflective tapes significantly reduced the grazing of mute swans especially when installed in criss-cross manner (Parrot and Watola, 2008). Reflective ribbons placed at 8-10 ft apart using bamboo poles evenly spaced above the trees and throughout the mango and citrus orchards have substantially minimized the parakeet, *P. krameri* attacks compared to acetylene exploders, reflecting mirrors and bird scaring models. However, the blackbird pillage remained persistent in the presence of reflecting ribbons possibly due to the non-impact of ribbons for them (Khan *et al.*, 2011). Among the different mechanical repellents tested, reflecting ribbons provided maximum protection in maize with only 0.47% damage of cobs and sunflower with 1.06% damage by parakeet, *P. krameri*. Application of reflective ribbons is useful in maize and sunflower fields to inhibit parakeet and other birds such as crows, sparrows and starling (Ahmad *et al.*, 2012). In a study made at Vivekananda Institute of Hill Agriculture (VPKAS), Almora reflective ribbons were found very effective against parakeets in maturing wheat whereas not effective against spotted munias, *Lonchura punctulata* in seed crop of radish and millets.

Many studies have found reflective tape to be ineffective. Reflecting tape was not effective for repelling starlings (*S. vulgaris*), robins (*Turdus migratorius*), house finches (*Carpodacus mexicanus*), mockingbirds (*M. polyglottos*), catbirds (*Dumetella carolinensis*) feeding on blueberry (Tobin *et al.*, 1988) and wild turkeys feeding on vineyards (Coates *et al.*, 2010). Conover and Dolbeer (1989) found that tapes with strands at 52 ft interval in cornfields did not reduce damage by red-winged blackbirds. In both cases the number and configuration of tape strands may have contributed to its ineffectiveness; leaving large spaces between rows of tapes allowed birds to avoid the tapes and enter the crop. Also, placing tapes along rows rather than perpendicular to them allowed birds' easier access along the rows. Failing to tape the whole field can also allow the birds an entry point into the crop. The 10 ft spacing was most effective for repelling blackbirds, however 16.5 and 23 ft intervals also provided some protection (Dolbeer *et al.*, 1986a). However, close rows of tapes throughout a complete field can increase the costs and even if bird damage is reduced the technique may not be cost-effective. The tapes are susceptible to damage by wind, entailing extra labour for repairs. Winds exceeding 15 mph frequently break the strands of tape or entangled them in trees or poles and thus not useful in windy places (Tipton *et al.*, 1989). Further, the metallic red colour is found to fade in one week period in days of bright sun shine.

Studies indicate that there may be a species-specific response to reflecting tape. Dolbeer *et al.* (1986a) speculated that tape may be most effective against flock-feeding birds, whereas those birds feeding solitarily or in small groups may be less sensitive to the visual stimuli. Reflecting tape may be most suited for protecting small fields of crops and gardens from certain depredating bird species (Bruggers *et al.*, 1986). In most trials, reflection tape reduced crop damage, where un-taped plots were available as alternative feeding sites. Gorenzel *et al.* (2003) reported that Mylar tape is effective against birds when first deployed but get habituated in just a few days so it is to be used along with other hazing techniques.

Although a close configuration of tapes may be successful in terms of crop protection, it can interfere with crop husbandry and increase costs in terms of labour and materials. In such situations, this technique is best suited to small areas of high value crops. Good maintenance of the tapes is essential in order to prevent them from becoming tangled in the crop and to stop gaps resulting from broken tapes being exploited as entry points by birds. Further work to develop this technique will be more promising by using a durable material. In general, tapes are useful for reducing bird numbers particularly if an alternative area for feeding is available.

1.4. Fire and Fire Imitators

Most animals and birds avoid fire. Since antiquity people use fire and noise for scaring crop raiding elephants (Osborn, 2002; Kalam, 2005; Sitati *et al.*, 2005). Both shouting and using fire in order to scare elephants away are traditional techniques, practiced for centuries (Osborn and Parker, 2003). Sitati *et al.* (2005) analyzed the methods of preventing elephant raids in 341 farms in Kenya, which revealed burning of fire as the most efficient method of all the traditional practices. The non-occurrence probability of elephant raids due to fire in Wald statistic is 31.7 followed by externally stationed guards (24.9) and banging tin and drums (4.26). In the past, villagers lit fires to scare away elephants, but now they are not allowed to do so (Sheil *et al.*, 2010) because of risk of accidents and sometimes elephants become aggressive. Moreover, fires at field boundaries or at elephant entry points to fields serve as a short-term deterrent, but are unsustainable for any length of time without large tracts of forest being cut down (Ngure, 1995).

Some materials are also burnt in fire to increase the deterrent effect of fire (Nelson *et al.*, 2003). People of Democratic Republic of Congo add capsicum seeds to fire to wade away elephants (Hillman-Smith *et al.*, 1995), while brickettes of elephant dung mixed with ground chillies are used in Zimbabwe (Hoare, 2001a; Osborn and Rasmussen, 1995). Farmers in the Waza-Logone, Cameroon believe that elephants dislike the smell of burnt sheep dung and use to burn it for elephant wading, but Tchamba (1996) found this to be ineffective. Flaming torches and hanging of kerosene lamps on wooden poles around the fields are said to be deterrent for crop raiding elephants (Nyhus *et al.*, 2000; Damiba and Ables, 1993). Fire and fire imitators make use of the common fear of animals to fire, causing them to flee from the area to be protected.

VL WAR- a Wild Animal Repellent

Considering the menace of wild animals in the region, VPKAS, Almora has developed an equipment which repels the nocturnal wild animals. The concept of burning fire for scaring and repelling the wild animals from the agricultural fields is well utilized here (Sushil *et al.*, 2009a). The major components of the equipment are funnel shaped lamp shade, fins, bottom with lock system, hanging silvery strips or reflective ribbons around the fins and CFL assembly. The equipment is made of galvanized iron sheet of 22 gauge. The roof of the equipment is inverted funnel on which a red plastic ball of ~150 mm diameter is fixed. Three fins of 450 mm length jointed at 120° with each other with lower portion having U-notch for accommodating the CFL. A chain of silvery strips of 22 mm width and 460 mm length is made by fixing the one end of the strips on a thin rope. Due to light weight of the strips, it starts flickering with a slight wind movement and imitates the fire. Field testing of the equipment, 'VL-WAR' was carried out by installing it in strategic locations in the most severely affected villages and reported that none of the crops were damaged in the visible range of the equipment during continuous three months of testing, otherwise, damage was observed every 3rd or 4th day in the same fields.

Depending on the topography, aspect of the mountain, path of the wild animals, VL-WAR has to hang about 1.5 to 2 m above the ground level. The equipment should be installed in the centre of the field or in front of the main path of entry of the wild animals. It is noteworthy, that most of the wild animals follow a fixed path for entering in the village or a particular area. The equipment should be operated throughout the night for scaring the nocturnal wild animals. It also serves the purpose of scarecrow during day time. Under hilly condition, normally it covers about one ha area (Sushil *et al.*, 2009b).

Photo 3: VL WAR- A wild animal repellent by imitating fire at night

1.5. Mirrors and Other Reflectors

The principle behind mirror use as bird deterrent is that they startle birds by producing a bright flash of light. Uhler and Creech (1939) described a spinner reflector used to frighten waterfowl from field crops. The device was made by attaching 12 inch square tin sheets on a horizontal wheel mounted on a 4 to 5 ft tall T-bar. Units were spaced 35 to 40 m apart throughout a field so that ducks could land no farther than 25 to 30 m from a reflector. Another device consists of aluminum pie plates suspended from varying lengths of line attached to an S- shaped rod suspended over the field. Wind and wave action caused the pie plates to move, which produced noise and light reflections. Whirling or flashing pieces of metal suspended in fields or on buildings reportedly have been used with limited effectiveness to scare birds (Frings and Frings, 1967).

Presence of mirrors during the breeding season has elicited aggressive behavior in glaucous-winged gulls, *Larus glaucescens* and female blue grouse, *Dendragapus obscurus* (Stout *et al.*, 1969; Stirling, 1968). When placed in nesting territories mirrors evoked aggressive responses from blue grouse, *D. obscurus* (Stirling, 1968). Foraging by black-capped chickadees, *Parus atricapillus* at feeding stations were depressed by keeping a standard mirror. Further, the dominant black-capped chickadees were more likely to get threatened with their mirror image than were mid-rank chickadees (Censky and Ficken, 1982). Seamans *et al.* (2001) evaluated the effectiveness of flashing lights combined with mirrors and mirrors alone as deterrents for starlings nesting in nest boxes. The treatments included placing mirrors on the back and two side walls of the nest boxes, mirrors with red flashing light and mirrors with green flashing lights, flat mirror on the back and convex mirror at the back and convex mirror above the entrance hole. Boxes with mirrors regardless of lights had fewer nests and fewer eggs, nestlings than control boxes. No difference was reported in fledglings produced per nest with nestlings of each treatment. After two years, nest boxes with complete mirror coverage showed the lowest occupancy rate. Thus mirrors are found to be repellents of starlings but not a practical method as such to repel starlings from their nests. Mirrors alone will probably not suffice as a repellent technique. The use of mirrors reflecting sunlight has failed to repel starlings, pigeons or gulls (Belton, 1976; Seamans *et al.*, 2001). In a blueberry field, fish crows (*Corvus ossifragus*) were observed feeding within 6 m of an operating mirrored device with red mirrors seven days after placement (Seamans *et al.*, 2003). Both red and clear mirrors did not appear to deter red-winged blackbirds from feeding in sweet corn fields. Ring-billed gulls, *L. delawarensis* were not deterred from roosting on an industrial roof by red or clear rotating mirrors. Although the gulls stayed about 5 m from the devices, they continued to use the roof (Seamans *et al.*, 2003).

A device consisting of a rotating pyramid of mirrors has been recommended for preventing crow damage to seedling corn (Bishop *et. al.*, 2003). A device containing three triangular mirrors (15x11.5 cm) positioned in a pyramid; in one set with clear mirrors and in the other with two red mirrors was evaluated for deterrence of pest birds. The device was powered by a 12-v battery, rotates at approx. 30 revolutions/ min and deflects sunlight with the intention of confusing birds, thus making sites unattractive to birds. Brown-headed cowbirds (*Molothrus ater*), common grackles (*Quiscalus quiscula*), European starling, house finch, red-winged blackbird, blue jay (*C. cristata*) and white-crowned sparrow (*Zonotrichia leucophrys*) were not found to be repelled by these mirrors (Seamans *et. al.*, 2003). Some cherry growers use lighting or shining metal or mirror piece as bird repellents (Simon, 2008). The multi-mirror reflectors, provided with an adjustable steel pipe equipped with an electric motor for their circular rotation at a speed of 7 rpm was used to deter predation of parakeets, *P. krameri* in mango, guava and citrus orchards and found less successful (Khan *et. al.*, 2011). The same kind of multi-mirror reflectors are found effective in preventing *P. krameri* from damaging maize cobs and sunflower seeds but next only to reflecting ribbons (Ahmad *et. al.*, 2012). The important disadvantage of this is habituation which generally occurs when reflectors are used for extended periods or over large areas

Reflectors (Scott and Townsend, 1985), tinsel, aluminum plates or pans (Hale, 1973), flashing, whirling strips or disks and pieces of tin (True, 1932) have also been used to protect small acreages of crops and gardens from animal damage. Suspended pieces of tin (Spalding, 1885) and whirling, twisting or fluttering strips or disks have been used to repel rabbits (Koehler *et al.*, 1990). Similarly, aluminum pie pans, tin can lids, plastic windmills, etc. have been explored to repel raccoons, opossums and skunks (*Spilogale* spp. and *Mephitis* spp.). Water filled clear glass bottles/jars or empty wine bottles planted upside down are kept in the periphery of the gardens to repel small mammals such as rabbits and groundhogs, *Marmota* spp. (Koehler *et al.*, 1990).

1.6. Flash Lights and Laser Lights

Flashing lights is a novel stimulus to birds, which encourage an avoidance response (Harris and Davis, 1998). Various types of lights (strobe, flashing, revolving and search) have been used in attempts to deter birds at feeding, loafing and roosting sites (Krzysik, 1987; Koski *et al.*, 1993). To effectively use light in managing birds, an understanding of avian vision is critical. Colour and type of light used to frighten birds have shown species-specific reactions ranging from indifference to flight (Belton, 1976; Blackwell, 2002; Gorenzel *et al.*, 2002). Many birds discriminate the colour of light at wavelengths between 400 and 700 nm (Pearson, 1972). In addition, some species, including pigeons, mallards (*Anas platyrhynchos*), belted kingfishers (*Megaceryle alcyon*) and some passerines perceive ultraviolet light (<390 nm) (Bowmaker and Martin, 1985; Martin, 1986; Cuthill *et al.*, 2000). Rock doves, *C. livia* and some songbirds have exhibited sensitivity to the plane of polarization of light (Able, 1982; Young and Martin, 1984), to which humans have very limited sensitivity.

In a study of the effects of strobe lights on laughing gulls and American kestrels (*Falco sparverius*), Bahr *et al.* (1992) found that strobe light frequencies of 50 Hz elicited faster responses in heart rates than frequencies of 5, 9 and 15 Hz. Frequencies from 8-12 Hz produce stress in gulls, pigeons and starlings (Solman, 1976; Belton, 1976). Similar studies were done by Green *et al.* (1994) through strobe lights of varying wavelengths (colour) and frequency against laughing gulls and American kestrels in laboratory. The test birds responded physiologically (increased heart rates) to strobe-light stimuli but avoidance reactions were not observed. A study reported that scare distance increased with an increase in strobe frequency but not exceeds to 100 Hz. A few studies using strobe lights, amber barricade lights and revolving lights on aquaculture facilities (Salmon *et al.*, 1986; Knittle and Parker, 1988; Littauer, 1990) indicate that these lights are effective in deterring night-feeding birds (e.g. herons). Belton (1976) found that gulls delayed approaching a feeding area by 30 to 45 min when it was illuminated by a white or magenta strobe at 2 Hz.

Strobes and searchlights have been used to deter birds with mixed level of success (Larkin *et al.*, 1975; Lawrence *et al.*, 1975). Birds such as kites, vultures and pigeons are affected by a high intensity strobe light, which could encourage them to take evasive action and move away. Strobe lights appeared to be more effective at deterring lapwings, *Vanellus* sp. than gulls (Harris and Davis, 1998). Uhler and Creech (1939) described an early homemade revolving beacon with reflectors they used to repel waterfowl from grain fields. The beam was reflected off the spinning reflectors, causing a series of flashes that effectively frightened ducks from the field throughout the harvest season. Blinking road-flasher lights were used by Lostetter (1960) to protect a 10 acre alfalfa field being damaged by widgeon, *Anas americanus* by placing just above the crop level. Searchlights have been used to deter ducks from landing and feeding in grain fields and tests have shown that some nocturnal migrants illuminated by light beams take evasive action.

Lights have been used with poor to partial success on heron, gulls and waterfowl (Gorenzel and Salmon, 2008). Pritts (2001) surrounded a blueberry planting with strobe lights, but found not effective for repelling birds. Belant *et al.* (1997b) tested flashing lights in starling nest boxes and found them ineffective at deterring starlings from nesting in the boxes. Flashing lights were reported effective against starlings (Lefebvre and Mott, 1987) and various light emitting devices are recommended as short-term deterrents for night-feeding bird predators also (Salmon *et al.*, 1986; Parkhurst *et al.*, 1987).

Badgers can be discouraged from areas to be protected by using bright lights (Johnson, 1983). A continuous three years survey in Assam revealed spotlights as an effective measure to prevent crop damage by elephants (Davies *et al.*, 2011). Villagers in Sumatra use powerful flashlights to deter elephants in combination with noise and fire (Nyhus *et al.*, 2000). The disadvantage is it cannot be used to drive away the animal to a distance far enough to prevent their return.

The effectiveness of revolving lights depends on their size, proper placement in the field, size of the field and number of units used (Lostetter, 1960). The lights probably produce a blinding effect so that birds become confused and cannot feed and thus try to fly away. Blind spots can occur if trees or other obstructions block the beams (Horn, 1949) and birds like black crowned night herons, *Nycticorax nycticorax* when tested landed to avoid bright light (Gorenzel and Salmon, 2008). In some cases, birds became habituated to the lights and even learned to avoid the lights by landing with their backs to them. During foggy or rainy weather, lights may attract birds instead of repelling them.

Laser lights

Lasers have been suggested as a technique for dispersing birds (Lustick, 1973; Lawrence *et al.*, 1975). Use of hand-held laser devices that project a 1 inch diameter red beam is used to disperse birds. These devises have been successful to disperse birds such as Canada geese, double-crested cormorants and crows from night time roosting areas in reservoirs and trees (Cleary and Dolbeer, 2005). The low power levels, accuracy over distance, silence and the ability to direct them on specific problem birds makes laser devices an attractive alternative to other avian scaring devices. Birds are startled by the strong contrast between the ambient light and the laser beam. It is found to be safe if it is not pointed at an unprotected eye within a distance of 155 m (McKay *et al.*, 1999). The use of lasers can be an effective method of bird scaring, although there is some evidence to suggest some birds are laser-resistant (McKay *et al.*, 1999). Laser light producing machines is like a pistol and the beam will be projected as long as the trigger is held down. Some devices even have a range of 20,000 ft. No ocular or other injury to the birds is reported from these lasers in double crested cormorants tested at a distance of 33, 13 and 1 m (Glahn, 2000). The electro-magnetic energy associated with lasers can cause stress, discomfort and behavioural effects in both birds and mammals including humans. If the energy is powerful enough, heating and physical damage can also occur. The hypothesis is that birds would leave areas where they were disturbed in this manner.

Lasers have been tested in a number of studies with mixed results. Seubert (1965) described experiments in which caged gulls were exposed to pulsing lasers. Pulsed light at low powers (1-2 joules) produced some flinching but no distress or alarm calls. Light pulses of 100-200 joules directed at the birds singed feathers and caused bleeding in the bird's eyes. However, the gulls did not react to lights stronger than to the 1-2 joule light. A continuous laser was also tested but the gulls looked directly into the beam of intense red light with no appearance of discomfort. Mossler (1980) tested whether the beam from a helium-neon laser would deter gulls at a landfill from feeding on highly-attractive food. The gulls showed some limited behavioural reactions to the laser beam but it did not deter them from feeding.

McKay *et al.* (1999) used a laser on two cormorant roost sites, both of which were disturbed. Although, tests are still required for many species, cormorants and Canada geese have shown extreme avoidance to laser beams in field trials. Other species that react to laser beams include wading birds, gulls, crows and vultures. Tests on flightless geese indicated they could be herded in a desired direction with the laser beam (Gorenzel and Salmon, 2008). Blackwell *et al.* (2002) reported no effect on starlings, *S. vulgaris*, limited effect on rock doves, *C. livia* and some success in geese and mallards, *A. platyrhynchos*. However they also got habituated in 20 min. Gorenzel *et al.* (2002) found similar results with American crows, *C. brachyrhynchos*. Most crows were dispersed from roosts by the laser, but returned within 15 min. In three experiments with stationary and moving laser beams treating a randomly selected perch, brown-headed cowbirds were not found to be repelled. Similarly, a moving beam did not repel European starlings from treated perches, nor were they dispersed when targeted. Rock doves exhibited avoidance behaviour only during the first 5 min of usage (Blackwell *et al.*, 2002).

Laser lights are reported to be effective for some birds and not for others so studies are needed to evaluate species-specific responses relative to laser power, beam type, wavelength, light conditions and captive versus field scenarios. The advantage of laser lights is because of its silent operation, it proves good for locations where noise is a concern. They are found effective at long range of over ¼ mile (Cleary and Dolbeer, 2005). However, these lights may not be effective during daylight hours and their ability to scare birds at night varies with the bird species. Although experiments suggested that starlings, mallards and herring gulls were disturbed by either pulsed or continuous laser light (Lustick, 1973), the light had to be directed at sensitive areas on the birds. When aimed at the feathers, birds did not react even though the laser was capable of igniting their feathers. Green and blue lasers were ineffective at deterring white-tailed deer, *O. virginianus* (VerCauteren *et al.*, 2006). These lasers are not yet been tested for efficacy in repelling birds. However, qualitative evidence at some airports suggests green lasers can be highly effective at dispersing birds such as rock doves and European starlings. As a general safety precaution, lasers are recommended not to be pointed at people within a nominal ocular hazard distance of 43ft.

1.7. Scarecrows/ Human Effigies

Scarecrows work by mimicking natural predators, often humans, thereby causing birds and animals to flee the site (Harris and Davis, 1998). Scarecrows have a long history of use against pest birds (Frings and Frings, 1967; Achiron, 1968). Most scarecrows are human-shaped effigies; constructed from a wide variety of inexpensive materials such as grain sacks or old clothes stuffed with straw. The use of traditional scarecrows to deter grain-eating birds has provided variable success. A mannequin tested on rufous turtle doves, *S. orientalis* in a flight cage was found to have a larger effective area than a stuffed crow or kite (Nakamura, 1997). Doves and pigeons show strong response to human scarecrow than other effigies (Yui, 1988; Shimizu *et al.*, 1988) although the intensity of response varies depending on the existence of alternate food (Watanabe *et al.*, 1988). Simple scarecrows made of black plastic bags attached to wooden stakes are used to deter waterfowl from grain fields (Knittle and Porter, 1988). Boag and Lewin (1980) found that a human effigy was effective in deterring dabbling and diving ducks from small natural ponds. When the effigy was present, the total number of ducks on the ponds was reportedly declined by 95%. A scarecrow mounted on a float was 80% effective in deterring birds from circular ponds, but kingfishers were not repelled (Lagler, 1939). One of 14 fish-rearing facilities surveyed by Parkhurst *et al.* (1987) reported successful bird control with scarecrows, whereas 13 facilities rated them of limited or no success. Often, the traditional motionless scarecrows provide only short-term protection or are ineffective. Some birds may even utilize them as perches (DeHaven, 1971) or associate them with favourable conditions (Inglis, 1980). Traditional scarecrows did not give good success in deterring blackbirds from rice fields (DeHaven, 1971).

A variety of scarecrow models has been tested against various birds. One promising model consists of a three dimensional human effigy whose head and outstretched arms move periodically. The movement presumably more realistically mimics an alarming situation than does an unanimated model (Inglis, 1980). Pop-up scarecrow units that work in synchrony with propane exploders also have been developed and evaluated in agricultural fields. One version consists of a head and torso of a human effigy mounted on an exploder (Achiron, 1988). When the exploder blasts, the effigy shoots into the air and spirals back down with fringes fluttering from its outstretched arms. Pop-up scarecrows were found effective for deterring blackbirds from sunflower but they were less effective in fields where birds had an established feeding pattern and in fields located near roosts (Marsh *et al.*, 1991).

Scarecrows are more effective if they are moved every 2-3 days (Hussain, 1990). Scarecrows that move in the wind and that are used with other deterrent devices are more effective than immobile scarecrows that are not used with complementary devices. Littauer (1990) suggested that periodically driving a vehicle near the scarecrow or placing the scarecrow on a stationary vehicle could increase its effectiveness. A mobile scarecrow unit which consists of an inflated human effigy placed on a 3 wheeled cart that is guided along cables in fields and orchards are found effective (Achiron, 1988). Nomsen (1989) reported that a human-like scarecrow that popped up from a double propane cannon when it fired was very successful in keeping blackbirds from feeding over 4-6 acres of sunflowers. Ducks and geese were observed to be much easier to frighten off than blackbirds.

Cummings *et al.* (1986) used a propane cannon and a CO_2 pop-up scarecrow to deter blackbirds from sunflowers. They found that most birds were frightened away by the scarecrows; fewer birds returned during the treatment period than were observed during the control period. They speculated that the birds that returned had become habituated to the scarecrow. Inflatable, human-shaped bag that is mounted on a battery-powered compressor or electric fan which inflates every five minutes was found to repel great blue herons, *Ardea herodias* and black crowned night herons, *N. nycticorax* initially from a hatchery but found to habituate quickly. The numbers increased after the first four nights (Andelt *et al.*, 1997). Another model of scarecrow consists of an inflatable, human-shaped bag that is mounted on a battery-powered compressor or electric fan, which inflates every five minutes. Timers can also be connected to a photo cell switch which would allow the scarecrow-inflation sequence to begin at dusk or dawn. Once inflated, the scarecrow stands up and then emits a screeching, siren-like noise before it deflates (Littauer, 1990). The use of the Scary Man® devices to deter cormorants from ponds resulted in an immediate, drastic decrease for first 7 days of the treatment period and it has increased after 30 days of treatment. The enhancement of the scary man deterrent by placing hats and camouflage masks and propped them up to make them more closely resemble a shooter squatting also had some effect on the repulsion of the bird (Stickley and King, 1993).

An animal activated scarecrow was found effective for about 6 weeks to deter white-tailed deer, *O. virginianus* in soybean fields (Beringer *et al.*, 2003). The scarecrow used is portable and charged biweekly using batteries and air recharged at weekly intervals and the activation is through IR sensors. Real-time and video observations and deer tracks around the plots suggested that deer attempted to enter soybean plots almost immediately after the scarecrow was installed. Subsequent attempts to enter the fields did not occur for several days. In most instances deer fled from treated fields when the scarecrow was activated. The animal activated scarecrow was reported to be better than propane exploders (Belant *et al.*, 1996) and similar in protection as crop protection dogs (Beringer *et al.*, 1994). Human effigies/ scarecrows are used in places frequented by elephant raids, but elephants quickly become habituated (Hoare, 2001b).

Photo 4: Reflecting ribbons in barnyard millet field

Photo 5: Spotted munias feeding in fields with reflecting ribbons

Photo 6: Reflecting surface illuminated with light to deter animals at night

Photo 7: Scare crow used in agricultural fields

Photo 8: Keeping dead bird to deter birds in rice field

Photo 9: Beating tins to wade off birds from millet fields

Scarecrows need to be highly visible, life-like and moved on a regular basis in order to retain their effectiveness (Vaudry, 1979). The greater the realism, in appearance and behaviour of the scarecrow, the greater is its effectiveness. Because the threat that birds might associate with scarecrows is perceived rather than real. Habituation is likely to occur relatively quickly unless other scare techniques are used in conjunction. Brightly coloured loose clothing may help increase the effectiveness of scarecrow because they move in the wind and birds react more readily to coloured and moving objects. Painting scarecrows a bright colour can increase their detestability (Littauer, 1990). Scarecrows can be fitted with bright streamers that move and create noise in the wind effectively becoming a moving visual (Vaudry, 1979).

1.8. Dead Decoys

The visual deterrent that has been used successfully for birds is the display of dead birds in a 'death pose'. Realistic dead bird effigies of gulls and turkey vultures have shown promise as species-specific frightening devices (Saul, 1967; Stout and Schuzab, 1979; Seamans *et al.*, 2000; Avery *et al.*, 2002; Tillman *et al.*, 2002). Initial trials using dead gulls and ravens suspended from poles have also shown promising results in dispersing these species from feeding and resting sites. Several experiments and field demonstrations have shown that a dead turkey vulture (freeze-dried taxidermy mount with wings spread), hung by its feet in a vulture roosting or perching area, cause vultures to abandon the site. The point here is the dead bird must be hung in a 'death pose' to be effective. Dead birds lying supine on the ground or in the roost are generally ignored or might even attract other birds (Cleary and Dolbeer, 2005).

A supine and hanging turkey vulture effigy (a taxidermy mount) was found to disperse roosts of black vultures, *Coragyps atratus* and turkey vultures, *Cathartes aura*. In all tests, fewer vultures were observed in the roost during the treatment period when compared to the pretreatment period. Vulture effigies are promising tools that may be used as part of integrated programs to disperse vultures from problem roosting sites (Seamans, 2004). The effigy heads are made from turkey decoy heads and were painted the proper colours and wired to the body by the taxidermist. Real heads tend to disintegrate and shrivel, while fake decoy heads keep their colour and open eyes, thus adding to the overall scaring effect. The effigies seem to be equally effective if hung inverted with wings fully opened, partially closed or fully closed. Within two days of deployment of effigies, vultures were reported to get dispersed and joined another roost roughly 8 km away. After 2 to 4 weeks of removal of effigies the vultures returned to the sites, redeployed the effigies and again the vultures moved away temporarily, returning after a brief absence of effigies. It became clear that the effigies at roost sites needed to stay up indefinitely. On a 30 m cell phone tower, hanging the effigy at the bottom, below 30 m in height, seemed almost as effective as hanging it at the top (Ball, 2009). Avery *et al.* (2002) found that the effectiveness was maintained 5 months post treatment.

Models of dead gulls or actual dead gulls displayed prominently have been useful in scaring gulls away from some airports (Howard, 1992). Stuffed gulls in injured positions were found to be effective in dispersing gulls if they were frequently moved to avoid habituation (Hardenberg, 1965). Crucified corpses and polystyrene gull models were successful at keeping gulls off loafing areas but found less successful if alternate loafing areas are not found (Caithness, 1970). Ring-billed gulls, *L. delawarensis* and herring gulls, *L. argentatus* effigy was hung in a head down position the wings extended down beyond the head and tested in different places like land fill, nesting site, air port etc. On day one of effigy placement, no gulls landed on the active face in the land fill but about 400 circled above the active face. On days 2 to 5, the gulls circled and then landed on the active face amongst the effigies. Gull reactions to a hanging gull effigy differed based on the situation in which the gulls confronted the effigy. When the effigy was in a location that was highly desired by gulls and there were no alternative areas readily available, gulls used the site in the presence of effigies. This was especially noticeable at nest sites and single source food sites. However, when effigies were placed in desired loafing areas, away from food sources, gulls used alternative areas and refrained from using loafing areas with effigies on them for an extended time (Seamans *et al.*, 2007a). Cattle egrets (*Bubulcus ibis*) shot dead and placed in clear view around the roost repelled other egrets and prevented other birds from landing (Fellows and Paton, 1988).

Canada goose can be hazed by using dead goose decoys (Mott and Timbrook, 1988). A device called the 'Dead Goose Decoy' is marketed as a non-lethal method to scare geese away from designated areas. This device consists of a plastic Canada goose decoy that has the form and appearance of a dead goose. The presence of goose effigies had an initial repellent effect at all sites tested. Canada geese were observed flying towards treated areas and then flaring away. By the end of the first 5 day treatment period, geese were generally returning to the field but were staying at least 25 m away from the effigies. Next one to three days, geese were found very near to the effigies (Seamans and Bernhardt, 2004). Decoys have been shown in several studies to have a significant deterrent effect on woodpigeons (Hunter, 1974; Inglis and Isaacson, 1984). Inglis and Isaacson (1984) confirm that real woodpigeon decoys, with open wings displaying the white wing marks had a significant effect on whether arriving woodpigeons would land or not. Indeed the wing alone as long as they have the white marks, deterred woodpigeons from landing. Hunter (1974) compared painted metal models of woodpigeons with real woodpigeon decoys and found that painted metal models of woodpigeons with wings extended to display the white wing marks, reduced damage to cabbages for about four weeks, after which the flock became habituated to the models. Crude woodpigeon silhouettes were not effective repellents, but real woodpigeon bodies with outstretched wings gave significant protection from bird damage over a nine week period. Pairs of wings and three-dimensional lifelike models were just as effective as corpses. However, corpses and pairs of wings had to be in good condition, so that visual clues are clearly visible to remain effective (Inglis and Isaacson, 1987).

The inefficiency of dead effigies by some birds species were also reported such as carrion crows, *Corvus corone corone* were not found to show avoidance to the hanging of dead carrion crows in a field of sprouting corn (Naef-Daenzer, 1983). Canada goose, *B. canadensis* effigies also did not show promise as effective goose control devices (Seamans and Bernhardt, 2004). Although this technique is inexpensive, its effectiveness varies under different circumstances and depends on, for example, frequently moving models (to prevent habituation), the availability of alternative sites (to which birds can relocate) or reinforcement with additional deterrent techniques such as pyrotechnics and alarm/ distress calls (Bishop *et al.*, 2003).

Chapter 2
Auditory Scarers

Sound is one form of communication used for territorial defense, mate choice, navigation, song learning of individuals and predator avoidance (Gill, 1995). Animals often have very acute and sensitive hearing, thus acoustic frightening devices may discourage animals from an area. Animals tend to initially avoid areas with loud and or unfamiliar sounds (Koehler *et al.*, 1990). Loud noises including explosion from gas exploders and sirens are commonly used as frightening devices (Gilsdorf *et al.*, 2002). Bioacoustics is animal communication signals often in the form of alarm or distress calls. An alarm call is a vocalization used to warn other individuals of possible danger and a distress call is emitted with an animal is being physically traumatized or restrained. Broadcasting alarm or distress calls of birds frighten them from the area to be protected (Bomford and O'Brien, 1990). Speculated advantages of using alarm /distress calls are that they are meaningful to the animals at relatively low intensities and are often species-specific reducing the non target effects.

2.1. Sound Making Instruments

Loud voice or sound which is not familiar will cause every living organism to startle and make them to run off. Human beings chasing animals with loud voices, screaming and beating drums or metal discs dates back to early ages. Many people believe, when they shoot at animals, that the animals flee because they fear death. It seems highly doubtful, from what we know of the conceptual abilities of birds and most mammals, that they can fear an abstraction such as death. The animals are undoubtedly frightened by the noise (Frings, 1964). Since this is the case, one is inclined to wonder why it has taken so long for firecrackers and guns to be discarded in favour of safer and cheaper mechanical noise generators.

Traditionally people use light, fire and noise to deter crop damaging elephants (Kalam, 2005). Beating tin cans, hurling rocks and bursting firecrackers are the primary methods used to scare off elephant raids (Madhusudan, 2003). Beating on drums or making a noise of any kind is one of the most common strategies to prevent elephant raids (Nelson *et al.*, 2003). Out of 79 farmers around the Maputo Elephant Reserve, Mozambique all used noise made by drumming on tins and pots to frighten off elephants, but only 52% confirmed this to be an effective method (DeBoer and Ntumi, 2001). Whip-cracking to imitate gunfire is used in both Africa and Asia (Hoare, 1995; Nyhus *et al.*, 2000). Burning of bamboo and causing it to 'explode' to wade off crop raiding elephants is practiced by communities in and around the Dzanga-Sangha Reserve in the Central African Republic (Kamiss and Turkalo, 1999). Cow bells are used to tie in ropes around the crop fields hoping it will deter animals especially elephants from entering the field (Graham and Ochieng, 2008). But elephants habituate to the sounds of people shouting, drums beating and guns firing (Osborn, 2002). Crop loss in cacao through squirrel (*Sciuridae* sp.), chimpanzee (*Pan troglodytes*), agile mangabey (*Cercocebus agilis*) and moustached guenon (*Cercopithecus cephus*) can be effectively managed by making noise such as beating metal drums (Arlet and Molleman, 2010). Beating on tin sheets or barrels with clubs also scares birds especially starlings (Johnson and Glahn, 1998) from agricultural fields. Two metal discs hung nearby in the fields so as to make sound by dashing while wind blows, frighten the pest birds to fly away. White noise is even comparable with distress call playbacks in some birds. Response of starling distress calls and white noise did not differ significantly but habituation was found to be quick to white noise (Johnson *et al.*, 1985). Noisemakers, including tin-can rattles and other rattling devices, vehicle horns or sirens and/or whistles have been used with variable success to repel or move such mammals as rabbits, deer, bison (Meagher, 1989) and coyotes (Koehler *et al.*, 1990).

The major drawback to the use of mere noises for chasing birds or mammals is that the animals stop responding after a time. Further, many studies revealed that animals which are acclimatized with the noises in the environment cannot be frightened even with loud noises. Birds and animals which are not exposed to noises expressed a significant difference in fearing to even low voices. Hens exposed to 90dB noise were reported to be more stressed and fearful than control hens (Campo *et al.*, 2005). Loud sounds is effective to frighten birds and animals from depredation of agricultural crops but cannot be used in places of livelihood since it disturbs human activities. Although these have been used successfully at some cattle feedlots, the loud noise may frighten some livestock also (Johnson and Glahn, 1998).

2.2. Crackers, Gun Fire and Fire Projectiles

The loud unnatural noises produced by crackers especially when exploded overhead, frighten most birds away from the source of the noise, at least temporarily. Firecrackers have reported to work temporarily even against carrion crows. Short-range shell crackers and other pyrotechnic devices were used with some success of deterring starlings (Seubert, 1964). Rope firecrackers have been used effectively against blackbirds in rice and corn fields, ducks in corn fields and fish-eating birds at fish hatcheries (Bivings, 1986). They have also been used to successfully disperse starlings from roosting sites and vineyards (Fitzwater, 1988). DeHaven (1971) indicated that about 5 acres of rice could be protected with an individual assembly of rope fire crackers that was elevated above the top of the rice panicles. No bird damage was reported to occur within 1,300 ft of an assembly of aerial bomb placed in a corn field under attack from blackbirds. An aerial bomb known as the 2-shot repeating bomb is manufactured especially for crop protection. Various types of rockets are also available for scaring birds and they are useful for frightening birds that are some distance from the operator (Vaudry, 1979).

Similar to fireworks, projectile devices rely on an explosion or other type of loud noise to deter birds from an area (Mott, 1980). Crackers produce visual stimuli such as a flash of light or burst of smoke. Flare pistols that shoot exploding or noisy projectiles, including shell crackers, bird bombs, bird whistles, whistle bombs or rocket bombs are used effectively in bird scaring (Booth, 1983). Bird bombs may travel 50 to 75 ft before it explodes whereas screamers or whistlers do not explode but makes screaming and whistling noises as they travel and leave a visible trail of smoke. These devices can be especially useful in situations where sites need only be protected for relatively short periods of time. Most bird species become habituated to these noises if used repeatedly over a long period of time. Gunfire is considered more effective over longer periods when supplemented with other frightening methods such as gas exploders, air horns, etc. (Hochbaum *et al.*, 1954; Dolbeer, 1980). Gunfire must be used with extreme caution because of the danger of stray bullets and exploding projectiles. The 0.22 caliber high speed, hollow point ammunition rifle is one of the best weapons known for frightening blackbirds from open fields of rice and corn (DeHaven, 1971; Meanly, 1971). From an elevated stand near the center of the field, one man with a 0.22 rifle has reportedly successfully kept blackbirds out of a 160 acre field (Neff and Meanley, 1957). Rifle fire alone can be effective in frightening waterfowl from rice fields, but its effectiveness is greatly improved when combined with a visual stimulus such as a scarecrow (Knittle and Porter, 1986). Besser (1985) recommends firing a round above the feeding or loafing birds, followed by a rapid series of shots behind them when they flush. Shooting to frighten jays from orchards is a common method used but not highly effective. Shotgun fires may not protect corn from blackbird attack very efficiently (Crabb *et al.*, 1986).

The exploding shotgun shells with firecrackers are more effective than the regular gun shots against blackbirds (Zajanc, 1962). DeCalesta and Hayes (1979) inferred that shell crackers repelled starlings from blueberry fields over 30 days without habituation. Whistle bombs, bird whistles and rocket bombs make hissing or whistling noises as they travel through the air but do not explode and thus are generally less effective than shell crackers or bird bombs, at least for some species (Booth, 1983). Bird bombs/ whistle bombs, if used intermittently along with cracker shells, provide a variation of sound that increases the effectiveness of both crackers and bird bombs to scare starlings, robins (*T. migratorius*) and cedar waxwings (*B. cedrorum*) from grape fields. Screamer shells are found very effective in dispersing flocks of Canada goose with no symptom of habituation (Aguilera *et al.*, 1991). Shell crackers in combination with taped distress calls were very effective (Marsh *et al.*, 1991). Brough (1967) performed a 12 month study scaring gulls, curvids, lapwings and starlings using warning calls and shell crackers. When shell crackers were used alone they were successful on 88 of 120 occasions (73%) and in combination it caused 93% success.

Firecrackers have worked temporarily against carrion crows (Suebert, 1964) whereas rope crackers are used extensively in bird damage programmes. The materials needed to make the rope firecracker are cotton rope, cotton twine and firecrackers. The cotton rope is cut to the desired length; the fuses of the firecrackers are inserted between the strands; and the rope assembly is suspended from the top with the twine. The cotton rope serves both as a support and a central fuse for the firecrackers which ignite as the rope burns from the lower end. The interval between explosions is determined by the burning rate of the rope and the spacing of the firecrackers. Burning speed of the rope can be influenced by its diameter and tightness of twist, its chemical treatment and weather conditions.

The advantages of pyrotechnics are these can be directed close to the bird/ animal and can influence the direction of dispersal. They are highly portable and can be used in various locations. Bird problem in agricultural lands can be reduced by removing the nearby roosts using fire crackers. This is demonstrated by Garner (1978) for the removal of blackbirds and starlings from the roosting sites. Persons were stationed around the roost at 300 to 400 ft intervals to fire aerial firecrackers in front of bird flocks as they approach the roost in the evening. The firecrackers were fired using an open bore 12 guage shotgun unless they are the shorter range special pistol type and shooting begun when the first flock tries to enter the roost and found successful.

Discharging firearms, cracker shells and/or other explosive shells are effectively used to repel deer (True, 1932; Scott and Townsend, 1985) and to direct bison movements (Meagher 1989). Gas exploders and various pyrotechnics are used to repel foxes, coyotes (Wade, 1983), bears (Lord, 1979), tree squirrels and rabbits as well as troublesome big game species such as deer, elk (*Cervus elaphus*) and pronghorn antelope (*Antilocapra americana*) (Koehler *et al.*, 1990). In Maharastra, farmers use various proactive techniques including explosion of firecrackers, whirling of flaming torches and the throwing of fire balls directly at the elephants. But it was found that elephants became accustomed to fire crackers and other scaring devices over a period of time and the techniques lost their effectiveness (Kulkarni *et al.*, 2008). Some projectiles may also prove to be a fire hazard as occasionally a malfunctioning round or smoldering debris of exploded bomb may ignite dry vegetation. So care should be taken while firing projectiles especially in dry seasons.

2.3. Propane Guns and Other Exploders

Gas-operated exploders, occasionally referred to as gas or propane cannons have been commonly used to repel pest birds in agriculture since the late 1940s (Achiron, 1968). Several types of exploders have been used effectively to reduce crop damage by blackbirds and starlings (Zajanc, 1962; DeGrazio, 1964; Stickley *et al.*, 1972). The unexpected bang produced by these exploders causes a 'startle' reflex and promotes escape flight (Harris and Davis, 1998). These devices produce extremely loud, intermittent explosions, usually at fixed 1 to 10 min intervals as desired. Present-day exploders consist of a bottled gas supply, separate pressure and combustion chambers, an igniting mechanism and a barrel to direct and intensify the noise of the explosion. The regulator at the gas supply can be manually adjusted to vary the interval between explosions, depending on the situation and bird species present.

Early gas exploders worked by igniting a mixture of air and acetylene produced by dripping water onto calcium carbide powder (Frings and Frings, 1967). Carbide exploders are used against Corvids and other birds, but their effectiveness generally lasts only for 2 to 6 days (Seubert, 1964) whereas they are very effective against prairie horned larks, *Eremophila alpestris* especially in damage prevention in newly planted crops (Koehler, 1962). However, these carbide exploders are sensitive to temperature changes and required daily maintenance to remain operational (Stephen, 1961; Zajanc, 1962). Carbide exploders were eventually replaced by more reliable models similar in action but operating with bottled acetylene gas, butane or propane. Today most exploders on the market operate with bottled propane gas. One of the newer models has been designed to operate electrically, which is governed/ directed by a solenoid valve releasing the gas into the exploding chamber where it is ignited by an electrical spark.

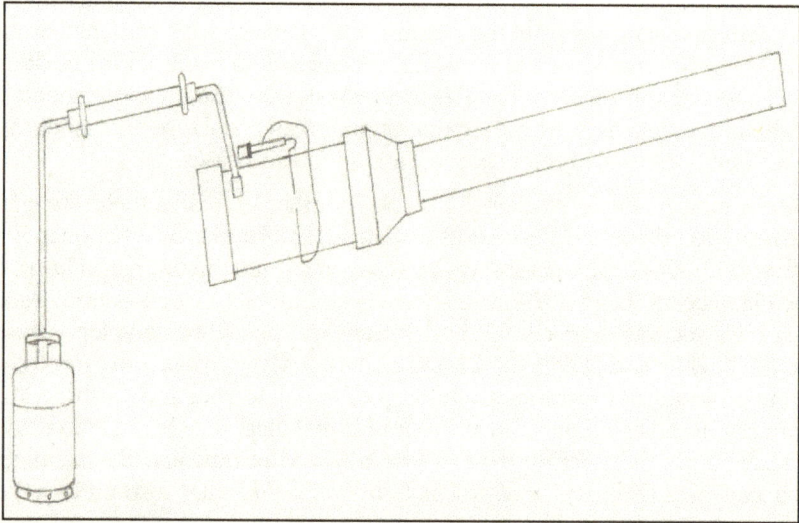

Figure 2.1: Propane gun

Exploders for bird pest management

Exploders have been useful for reducing blackbird damage to crops such as corn (DeGrazio, 1964; Dolbeer, 1980), rice (Pierce, 1972) and sunflower (Besser, 1978). Exploders were found to be the most cost-effective method for reducing blackbird damage, mostly by red-winged blackbirds, *A. phoeniceus* to ripening corn up to 77% when compared with untreated fields. Most damage occurred during the first 2 weeks, suggesting that habituation was not a factor (Conover, 1984). In agricultural fields with moderate bird pressure, one exploder can generally protect about 10 acres of crop from blackbirds (Pierce, 1972; Vaudry, 1979; Dolbeer, 1980); although DeHaven (1971) reported that up to 25 acres of rice can be protected if other devices are used along with the exploders (Marsh *et al.*, 1991). Besser (1985) found that exploders, when moved periodically within fields, provided adequate protection against blackbirds attacking sunflower during the 6 week period when the crop was susceptible to damage. DeGrazio (1964) reported that carbide exploders were used against blackbirds in corn field and the yield reduction from bird damage was only 1%, versus 43% in an unprotected field. Stickley *et al.* (1972) tested two exploders in each of six corn fields ranging from 5 to 17 acres and reported an 81% damage reduction by blackbirds during a 6 day test period whereas Avitrol, chemical toxicant reduced damage by 56%. Potvin and Bergeron (1981) reported that two non-synchronized exploders reduced blackbird damage in ripening field corn by 73%, whereas a single gas cannon, fired every two minutes offered not much protection. Acetylene exploders successfully reduced or stopped blackbird damage to ripening maize and reduced damage by 98% (Bishop *et al.*, 2003). Exploders have also been utilized to protect swathed wheat, barley and oat fields from waterfowl depredations and found green-winged teal, *Anas carolinensis* were more difficult to repel than were mallards, *A. platyrhynchos* and pintails, *A. acuta* (Stephen, 1961). Hobbs and Leon (1987) reported rotating exploders as a mechanical device to protect citrus groves from depredations of great-tailed grackles, *Quiscalus mexicanus*. Tipton *et al.* (1989) also used exploders as one of several methods to deter great-tailed grackles from citrus groves. Rappole *et al.* (1989) studied the detrimental effect of exploders on white-winged doves, *Zenaida asiatica* and mourning doves, *Z. macroura* nesting in the citrus groves and found birds were scared away from their nests up to 60 yards from the exploder. Propane guns were used in the harvest season in cherry orchards and found effective (Simon, 2008). But exploders apparently have not significantly reduced bird damage to grapes (Besser, 1985). Acetylene exploder proved to be the least effective in managing the parakeet, *P. krameri* attacks compared to reflective tapes and distress sounds in mango and citrus orchards whereas it was successful in guava orchards (Khan *et al.*, 2011). These exploders were least effective to prevent damage due to parakeet, *P. krameri* in maize and sunflower fields (Ahmad *et al.*, 2012).

The effectiveness of gas exploders depends on a variety of factors, including the species and number of birds present, availability of alternative sites for repelled birds, the density of exploders, interval between explosions and wind conditions (Blokpoel, 1976; Vaudry, 1979). A standard refillable 20 pound propane gas tank produces about 12,000 explosions. This would need replacement every two weeks at a firing rate of 13 explosions an hour or one explosion every four minute. Birds may become accustomed to the loud blasts after only a few days. To alleviate habituation, exploders should be moved periodically (every 1 to 3 days) within the area needing protection (Pierce, 1972; Salmon and Conte, 1981; Bivings, 1986). Stationary units can be elevated on a platform or tripod and faced downwind to increase their range (Besser, 1978; Kopp *et al.*, 1980). Mounting units on rotary tripods enables them to rotate after each blast, thereby projecting blasts in all directions; a feature intended to delay habituation. They are generally considered to work best when reinforced with other bird-frightening devices (DeGrazio, 1964; DeHaven, 1971; Kopp *et al.*, 1980).

Exploders for mammal pest management

Acoustic frightening devices like exploders have been recommended for deterring deer from desired areas (Koehler *et al.*, 1990; Craven and Hygnstrom, 1994). Propane exploders are used to repel numerous vertebrate pests, including deer (Craven and Hygnstrom, 1994).

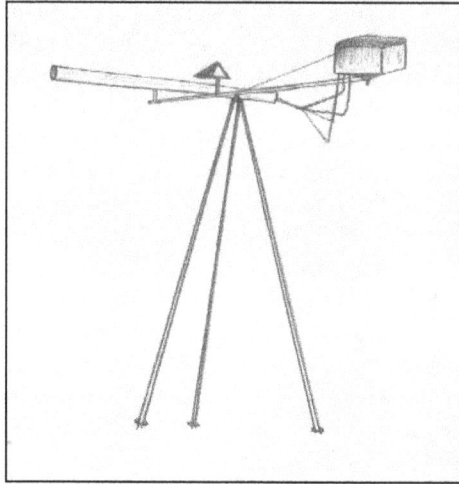

Figure. 2.3 Bird scaring gun

Exploders have generally been designed to detonate at standard time intervals (Stickley *et al.*, 1972; Cummings *et al.*, 1986), but are likely more effective when calibrated to detonate at random intervals (Bomfort and O'Brien, 1990). Belant *et al.* (1996) evaluated the effectiveness of propane exploders as deer deterrents in fields with high white-tailed deer density (91/ km²). Systematic exploders calibrated to detonate once at 8 to 10 min intervals and motion-activated exploders to detonate 8 times/deer intrusion were tested. Systematic propane exploders were generally ineffective, deterring deer from corn for <2 days only, whereas motion-activated exploders repelled deer up to 6 weeks. Repellency of motion activated exploders varied seasonally, possibly in response to variations in deer density, availability of alternate food or reproductive and social behavior.

As motion-activated exploders detonate only when the target species approaches the area to be protected, habituation may not occur as rapidly as with exploders activated at predetermined or random intervals. Operating exploders when damage first occurs, moving exploders frequently and elevating exploders to increase noise levels may also enhance their effectiveness (Craven and Hygnstrom, 1994).

Electronic Guards originally developed for frightening coyotes were reported effective for repelling deer only for less than one week period (Belant *et al.*, 1998a). An electronic guard and a propane exploder were tested against white tail deer damage in corn at the silking stage. No significant difference was obtained in yield or damage/ plot in any of the treatments with respect to control. Acoustic frightening devices had no effect on the area used by radio-marked deers and continued to use the cornfields that contained frightening devices (Gilsdorf *et al.*, 2004). So propane exploders were not much effective for mammals particularly when used in fixed place and fixed at regular intervals.

2.4. Sonics and Ultrasonics

Sonic devices broadcast sounds in the audible range of animals and birds which are intended to alarm or stress the target birds and either cause or predispose them to leave the area. In theory, any loud, startling noise will temporarily frighten birds (Frings and Frings, 1967) and animals (Bomford and O'Brien, 1990). Sounds are alleged to repel animals by several mechanisms: pain, fear, communication jamming, disorientation, audiogenic seizure, internal thermal effects, alarm or distress mimics etc. Ordinary loud noises have been used since antiquity to repel birds and mammals. These can be produced variously, from clapping hands to firing cannons. Devices occasionally used for hazing animal and bird pests include air horns, sirens etc (Theissen *et al.*, 1957; Parkhurst *et al.*, 1987). Sonic systems that produce a variety of electronically-produced sounds are also commercially available. They can emit noise at levels up to 120dB (Harris and Davis, 1998).

Air horns operate with compressed air to produce a loud, braying blast. Such units often are made up with a 12 volt air compressor and two trumpets to intensify the noise produced. It was reported that the starlings, *S. vulgaris* were kept away from 50 acres of ripe grapes in a vineyard for 7 weeks by broadcasting two different sounds (Seubert, 1964). Air hornes were used against starlings, house finch (*C. mexicanus*), mockingbird (*M. polyglottos*) and mourning dove (*Z. macroura*) in grape vineyard and found effective except for mourning doves (Marsh *et al.*, 1991). It is not much effective against crowned sparrows, *Zonotrichia* spp. in lettuce field also. A study of an acoustic device to reduce damage to grapes by grey-breasted white-eyes, *Zosterops lateralis* in Australia resulted in 83.3% reduction in damage in grapes compared to control plots without the device (Knight and Robinson, 1978). Theissen *et al.* (1957) reported that air siren has little practical value for hazing mallards, *Anas platyrhynchos* and pintails, *A. acuta* from agricultural fields. The distress sound players placed near the garden amplifying the fearsome sounds (explosions) to disperse the birds at frequent intervals did not found to reduce the attacks of parakeets, *P. krameri* to levels less than that of reflecting tapes in mango, citrus and guava orchards (Khan *et al.*, 2011).

Deer whistles are acoustic means to reduce deer-automobile collision by deterring them (Romin and Bissonette, 1996). A sonic device for producing sound was tested for deer deterring and reported effective for one week because the deer were affected by the novel stimuli. Habituation to these devices occurred rapidly within one week (Belant *et al.*, 1998a). This device has an advantage of being humane, inexpensive, easy to use and scientifically sound (Bomford and O'Brien, 1990). Six different whistles were tested in the laboratory and on motor vehicles against white-tailed deer, *O. virginianus*. The primary frequency of operation of the closed end whistles on vehicles was found to be 3.3 kHz while for the open-end whistles it was about 12 kHz. The best frequency range of hearing for white-tailed deer appears to be between 2 and 6 kHz. So the whistles of this range will be of more effective. Myers (1967) studied the response of rats in boxes to acoustic stimuli and found white noise as aversive at 85 dB and pure tones at 105 dB. Wilson and McKillop (1986) inferred that a sonic device was only mildly aversive to European rabbits, *Oryctolagus cuniculus* in a grass enclosure and lasted only for one week. In Laikipia, Kenya, cow bells have been hung around field of crops hoping that the noise of bells might alarm the elephants or at least alert the farmer to a potential crop-raiding event (Graham and Ochieng, 2008). Rai *et al.* (2009) reported an electronic security system for agriculture fields to combat animal damage. If an animal enters the field breaking the wire, the circuit sounds a loud alarm in a harsh tone which may compel the animal to run out of the field and at the same time alerting the farmer about the intrusion. This device is very much useful in protecting the field from animals.

Electric or air-produced nonspecific, audible loud sounds have limited potential for animal and bird hazing. Consequently the value of most electronic noises on the birds and mammals tested is almost limited to short term control. The best effects are obtained when (1) the sound is represented at random intervals; (2) a range of different sounds are used; (3) the sound source is moved frequently; (4) sounds are supported by additional methods, such as distress calls or visual devices; and (5) sounds are reinforced by real danger, such as shooting (Bomford and O'Brien, 1990). Because of these complexities, the success of sound deterrents is largely a function of the skill and motivation of the operator.

Ultrasonics

Short-wave, high frequencies above 20,000 Hz are known as ultrasound and are inaudible to people, although some animals such as dogs, bats and rodents can hear well into the ultrasonic range (Frings and Frings, 1967). Long-wave, low frequencies less than 20 Hz are called infrasound and are also inaudible to people. There is no evidence that ultrasound or infrasound have unique properties making them more likely to repel animals than audible sound (Bomford and O'Brien, 1990).

Though there are many commercially available ultrasonic bird repellents in the market, there is no evidence that ultrasonic devices deter birds. In fact, evidence indicates that most species of birds do not hear in the ultrasonic range (>20 kHz) (Harris and Davis, 1998) and so there is no biological basis for their use. Pigeons were undeterred by ultrasonic systems (Haag-Wackernagel, 2000; Woronecki, 1988). Equipment producing ultrasonic sound waves (20,000 cycles per second) had no visible effect on starlings feeding in a cattle feed-lot, nor did this frequency discourage English sparrows from nesting in a barn (Zajanc, 1962). Meylan (1978) tested an ultrasonic device on greenfinches and house and tree sparrows in sunflower crop and reported that it deterred birds from visiting the crop. Griffiths (1987) tested an ultrasound device on house finch (*C. mexicanus*), dark-eyed junco (*Junco hyemalis*), white-breasted chickadee (*Parus atricapillus*) and blue jay (*C. cristata*) and reported no significant effect on feeding time, seed consumption or numbers of birds visiting the sites.

Rats, mice and other small mammals can hear sounds that are ultrasonic for man, so acoustical stimuli for them could be inaudible to man. Animals can be killed by sound if the intensity is great enough. At an intensity of 1 watt/cm^2, sound kills insects almost immediately and destroys mice in about 30 sec. This may look promising, but it is really not. To kill the mouse it must be restrained in a small sound field. Obviously, if one can hold a mouse in a sound field for 30 sec, he could destroy it by a number of much cheaper methods. In short, high intensity sounds, even though they can kill or injure animals probably have little promise in practical pest control (Frings, 1964). The three devices which emit vibrations in the frequency range 0 to 20 kHz were tested against European moles, *Talpa europaea* and the vibrations of this frequency are

rapidly attenuated when passing through soil. Moles fitted with radio transmitters gave no indication of being repelled by the mechanical devices (Gorman and Lamb, 1994). High-frequency ultrasonic sounds are ineffective to repel European rabbits, *O. cuniculus* (Wilson and McKillop, 1986). Curtis *et al.* (1995) reported that the Super Yard Gard® ultrasonic device as ineffective as a deer deterrent. We cannot conclude from this because in this study sound was emitted at regular intervals rather than activated by movements of deer. Later Belant *et al.* (1998a) reported motion activated Yard Gard® as ineffective as deer repellent. Two such devices were tested on little brown bats, *Myotis lucifugus* and neither reported effective. Blackshaw *et al.* (1990) tested 7 ultrasonic devices on domestic dogs and found one device, with a signal of 120 dB at 1m and a frequency sweep between 5 and 55 kHz was inferred to be aversive.

Loud sounds, either audible or ultrasonic can kill rodents under laboratory conditions by increasing their body temperature (Danner *et al.*, 1964; Nelson and Seubert, 1966; Frings and Frings, 1967). These thermal effects would be difficult, expensive and perhaps dangerous to produce for use on free ranging animals and therefore seem unlikely to be a practical use in animal control. Many studies have rejected ultrasound as a practical means of rodent control (Kent and Grossman, 1968; Meehan, 1976; Lavoie and Glahn, 1977; Lund, 1984) and concluded that ultrasound either had no effect on target species, or had only a partial and transient effect.

2.5. Biosonics and Distress Calls

Biosonics or bioacoustics is biologically significant sound originated by animals. Biosonics as a repelling technique are based on acoustical signals emitted by birds and animals to convey information to conspecifics. Biosonic deterrents operate by broadcasting either distress calls of the target species or the calls of their natural predators. Distress calls are those emitted by them when being restrained, attacked by a predator or subjected to other types of severe conditions, whereas alarm or warning calls are usually given in response to the presence of an intruder or predator. The response is species specific, to some groups distress calls will make them fly away immediately and in some other groups such as gulls, alarm/distress calls initially act as an attractant with birds approaching the source, apparently to investigate, before flying away (Brough, 1968). Arousing social contexts including social separations or encounters with strangers can result in calls of increased emotional intensity as observed in rhesus monkeys, *Macaca mulatta* (Bayart *et al.*, 1990), red fronted lemurs, *Eulemur rufifrons* (Fichtel and Hammerschmidt, 2002), baboons, *Papio cynocephalus ursinus* (Rendall, 2003), guinea pigs, *Cavia porcellus* (Monticelli *et al.*, 2004), tree shrews, *Tupaia belangeri* (Schehka *et al.*, 2007) etc apart from a number of birds.

The first report on the potential of distress calls in repelling birds' way back to 1950s, (Frings and Jumber, 1954) in which the piercing distress calls of the European starlings, *S. vulgaris* held by their legs was reported to frighten other starlings. Recordings of these calls broadcast through a loudspeaker were used to disperse starlings from several urban roosts (Frings *et al.*, 1955). After that, audible bird warning stimuli, distress and alarm calls are being explored and/or used for acoustically repelling birds to keep them out of areas where they are causing problems (Pearson *et al.*, 1967; Brough, 1969). Warning calls have been used to deter starlings, gulls (*Larus* spp.), corvids (*Corvus* spp.),

Canada geese (*B. canadensis*), night herons (*N. nycticorax*) and others by broadcasting in short bursts (10 to 90 sec.) depending on the bird species and situation. Distress calls are reported to be highly effective to wade away starlings from vineyards (McCracken, 1972), blue berry fields (deCalesta and Hayes, 1979) but effective only for a week in cherry orchards (Summers, 1985). Alarm call broadcasts were also used successfully in wading away starlings from orchards and vineyards (Summers, 1985; Feare, 1989). Starlings were repelled from a 50 acre German vineyard for 7 weeks by broadcasting calls from 6 loud speakers (Nelson and Seubert, 1966). Boudreau (1975) stated that distress calls temporarily scares starlings and may have limited use for dispersing starling roosts, but it has little effect on other bird species.

Distress calls were reported to disturb blackbirds, causing them to leave the field, at least temporarily. The tape-recorded distress call of a young blackbird held by one wing when played and directed toward flocks of feeding birds, the flocks were frightened (Zajanc, 1962). Distress calls were found highly effective against red-winged blackbirds, *A. phoeniceus* and corvids, *Corvus* spp. in corn fields (Naef-Daenzer, 1983). It was reported that red-winged blackbirds were effectively repelled for 25 days from a corn field and numbers of blackbirds and starlings were reduced 91% at a 20 acre feedlot after 80 days of broadcasting blackbird distress calls (Pearson, 1967). Taped distress calls of the immature male yellow-headed blackbird (*X. xanthocephalus*), the immature male red-winged blackbird and the adult female redwing were broadcasted on three corn fields having a history of heavy bird damage. Damage was reduced by 85, 23 and 15% on the three fields (DeGrazio, 1964). Keidar *et al.* (1975) found that distress calls deterred flocks of Skylarks, *Alaunda* sp. and Calandra larks, *Melanocorypha calandra* from feeding on agricultural crops.

Tape recorded crow distress/alarm calls were reported very effective in dispersing crows from urban roosts (Gorenzel and Salmon, 1993). Broadcast distress calls have been shown to effectively haze crows out of orchards (Salmon *et al.*, 1999). American crow (*C. brachyrhynchos*), a major pest in almonds, was effectively hazed out of almond orchards with broadcast of distress calls. It is recommended at a rate of one unit per 1.6 ha, when the first sign of damage is seen and distributed uniformly throughout the orchard, moved to a new tree every two weeks and automatically switched to a different call every three to four days. The important point is to start broadcasting at the first sign of bird damage is seen. The damage reduction in one experimental site was reported from 0.84 (6.0 kg/ha) to 0.25 (1.1 kg/ha) and from 1.54 (18.2 kg/ha) to 0.73 (4.8 kg/ha) in another site (Houk *et al.*, 2004). Depending on the species and situation, these warning calls often cause the conspecifics and sometimes closely related species to leave the immediate area. Rooks (*Corvus frugilegus*), jackdaws (*C. monedula*) and carrion crows (*C. corone*) could all be repelled from agricultural fields for up to 2 weeks by broadcasting distress calls of anyone species. Pear orchards were protected by using distress calls against those corvids for about two months (Seubert, 1964; Nelson and Seubert, 1966). Sometimes, distress calls are very specific to the locality. Distress calls of Eastern crow did not have deterring action for their European colleagues. This was also reported in Herring Gull (*L. argentatus*), wherein the birds in Europe is reported to have one set of signals and in America, a different set. This means that recordings may not be generally valid for a species; instead one may have to record calls for local use only.

Boudreau (1968) used distress calls to repel house sparrows, *P. domesticus* from a millet field. Martin (1976) tested the biosonic, Av-Alarm® against golden sparrow, *Passer luteus* and house sparrow, *P. domesticus* and found ineffective since it is effective up to 4 days only. Holcomb (1977), in contrast, found that the Av-Alarm® as effective in reducing damage to rice by red-billed quelea, *Q. quelea*. Distress calls of the house sparrow had some effect at deterring feeding sparrows from 2 grain fields, but crop damage was reduced only in a limited area (0.25 acres) around the loudspeaker (Bridgman, 1976). deCalesta and Hayes (1979) had only limited success in repelling cedar waxwings, *B. cedrorum* from 4 blueberry fields. They also reported that robins, *T. migratorius* did not respond to the starling distress calls.

Swallow sound files (8 alarm and distress calls) were selected and broadcasted to deter cliff swallow from perching and nesting in the bridges. Files contained multiple individuals within a colony giving alarm calls. The broadcast units were turned on at sunrise silent for 1 min and played all calls with 2 sec of silence between calls. These bio-acoustic broadcasts were found to reduce nesting of swallows by 53% (Conklin *et al.*, 2009).

The success with the starling and blackbird distress calls led to the study of distress calls of Herring Gull, *L. argentatus* but the later was found not to produce one. At the same time an alarm call, which is produced when a gull sights a captive or dead gull is found very repellent, if broadcast correctly (Frings, 1964). Bridgman (1976) reported that resting and feeding gulls react to broadcasts of distress calls of their own or closely related species by getting up and approaching the sound source. They then circle overhead for a while before drifting away in various directions to some other favoured site. Rarely do they attempt to resettle on or near the original location. Crows react in a similar manner, as noted by Salmon *et al.* (2000). Swamy *et al.* (1980) reported that distress calls are highly effective against Indian baya, *Ploceus philippinus* from roosting. Distress calls were also studied in linnets (*Carduelis cannabina*), English sparrows, prairie horned larks (*Eremophila alpestris praticolas*), robins and some species of doves (Frings, 1964). Four starling (*S. vulgaris*) calls, three finch (*C. mexicanus*) calls and one robin (*T. migratorius*) call were broadcasted in vineyard to deter the bird pests. The units were deployed at a density of 0.6 ha per broadcast unit, concentrated on the perimeter of the vineyard and moved weekly in a fixed pattern and found to be effective (Berge *et al.*, 2007). Naef-Daenzer (1983) performed a study on the scaring of carrion crows, *C. corone* with distress calls and suspended bodies of dead crows. It was played from dawn until dusk, each call was 20-30 sec and each silent period was 25 min. The fields treated with distress calls had significantly less damage than the fields treated with the dead crows and the untreated fields. There was no habituation to the calls at the reported calling rate.

The crimson rosellas (*Platycercus elegans*), a pest of several commercial fruit crops is considerable to have several acoustic variations. A study was made with three types of bioacoustic stimuli on *P. elegans* in apple orchards *viz.*, control stimuli, local alarm calls and non-local alarm calls which showed that the playbacks as very effective in reducing the activity of rosellas. Further no difference was found between the use of local and non-local alarm calls, which suggest that playback of alarm calls may be an effective deterrent of rosellas over a broad distribution, at least for short to medium-term use (Ribot *et al.*, 2011). Altogether, 9 wild turkey calls, 6 domestic turkey calls, 1 turkey putt voice call by a human, 4 crow calls and 2 calls with canine barks were tested to haze the wild turkeys, *Meleagris gallopavo* feeding on grapevines, but none found much effective. No significant difference was found in treated and control field was also found indicating the calls are ineffective in hazing the birds (Coates *et al.*, 2010).

Tests with distress calls broadcasted from a low-flying aircraft were initially effective in moving birds. However, once birds established feeding habits in a field, almost constant harassment was necessary for protection. Limited tests with distress calls broadcasted from the mobile sound unit revealed that the sound must be directional for maximum success. In these tests the maximum effective distance varied from 100 to 330m (DeGrazio, 1964). Aguilera *et al.* (1991) found that Canada geese in parks reacted to alarm/distress calls by becoming alert and sometimes moving up to 100 m away from the source of the call, but the birds did not leave the area. No doubt, distress calls have the advantage of being more effective than the use of unnatural sound and noises to repel animals (Marsh *et al.*, 1991) but not all animals make distress calls and these calls are most commonly emitted by gregarious species and large flocks usually are more responsive than small flocks or individuals (Brough, 1968; Boudreau, 1972). Distress calls have been used to disperse flocking birds like crows, starlings and gulls (Gorenzel and Salmon, 2008). Factors that likely influence their effectiveness include weather conditions, season, availability of alternative sites for repelled birds, group size, quality of recordings, number of broadcasting systems and possibly others (Blokpoel, 1976 ; Johnson *et al.*, 1985). Baxter (2000) performed a study using distress calls to deter birds at a frequency of not more than one 90 sec call during each half hour. The deterrence was successful in stopping birds from loafing and feeding on the site during the initial period, but after four to six weeks, habituation began to occur. The number of birds returned to pre-trial levels after 10 weeks of playing the calls. Harris and Davis (1998) suggested that raptor calls might be useful to deter birds since they might indicate that predators are close-by. However there is no properly controlled evidence to suggest this might work in practice, and since some raptors tend to hunt in silence, its biological relevance may be limited. But broadcasts of the protest calls of the sparrow hawk (*Accipiter* spp.) successfully repelled house sparrows and habituation was not observed after 6 days of exposure to the sounds (Frings and Frings, 1967). The playback of a Peregrine falcon (*F. peregrinus*) call was reported to be effective in dispersing gulls (Gunn, 1973).

Distress calls were also used against flying foxes, *Pteropus* sp. and reported successfully repelled them from orchards (Hall and Richards, 1987). Distress call records of rats being fed by skunk were played back on wild rats and found they are very repulsive and thus can be inferred that these show promise as a means of rat controlling mechanism (Sprock *et al.*, 1967) but found stressful to people too (Lund, 1975). Wade (1983) reported that recorded distress calls have been only temporarily effective against coyotes. In many cases, mammalian alarm calls vary acoustically according to specific predator species or class of predator (*e.g.*, aerial versus terrestrial). Experiments with the Suricates, *Suricata suricatta* (Manser, 2001) and vervet monkeys, *Cercopithecus aethiops* (Seyfarth and Cheney, 2003), showed that the listeners react to play back alarm calls as if they were in the presence of an actual predator. The complexity and variation of the acoustic cues can be seen in examples taken from three species of *Cercopithecus*, in which vervet monkeys, *C. aethiops* separate their alarm calls for leopards and eagles through the location of dominant frequencies (Seyfarth *et al.*, 1980), Campbell's monkeys, *C. campbelli* separate them by call duration, fundamental frequency and dominant frequency location (Zuberbuhler, 2001), while Diana monkeys *C. diana* separate them by call rate, duration, fundamental frequency and formant frequency location (Zuberbuhler *et al.*, 1997; Zuberbuhler, 2000; Riede and Zuberbuhler, 2003). Animal alarm calls are not always predator specific, however. For example, yellow-bellied marmot, *Marmota flaviventris*, alarm calls are similar towards a range of predators but do increase in rate with level of perceived risk (Blumstein and Armitage, 1997). African elephants, *Loxodonta africana* react similarly to sound playbacks of unfamiliar conspecifics (McComb *et al.*, 2001).

In Kenya, playbacks of elephant rumbles (either musth or oestrous rumbles) can have the effect of attracting elephants towards or repelling elephants away from the sound-emitting speaker (Poole, 1999). King *et al.* (2007) reported that the sound of disturbed African honeybees, *Apis mellifera scutellata* causes African elephants, *L. africana* to retreat. A significant majority of elephants, in a sample of 17 families and subgroups of varying sizes, reacted negatively *i.e.*, immediately walking or running away when they heard the buzz of disturbed bees, while they ignored the control sound of natural white-noise. Of the 17 families, 16 (94%) left the tree under which they had been resting within 80 sec of bee sound onset. For mode of retreat, 41.2% of families responding to bees 'ran' away, 29.4% 'walked fast' away and 23.5% 'walked' away. This moving away behaviour was not observed for the control groups. But signs of habituation start to appear within just two playbacks for elephant families who are subjected to playbacks with a short time gap between trials but elephants with a long time gap between trials show less habituation behaviour. Most importantly, elephants produced distinctive "rumble" vocalizations in response to bee sounds. Audio playbacks of these rumbles produced in response to bees elicited increased headshaking and further and faster retreat behavior in other elephants, compared to control rumble playbacks with lower second formant frequencies. These responses to the bee rumble stimuli occurred in the absence of any bees or bee sounds, suggesting that these elephant rumbles may function as referential signals, in which a formant frequency shift alerts nearby elephants about an external threat, in this case, the threat of bees (King *et al.*, 2010). This is concluded that the bee sounds alone will not be enough to deter elephants for long within one crop-raiding season, bee sounds can be deployed along with other deterrents with at least a five week time gap between each playback experience (King *et al.*, 2010). The playback of the call of a predator signals that a predator is nearby and animals and birds may react to this with heightened awareness. In natural situations, predators usually hunt silently so that they do not 'announce' their presence. Thus, the playback of predator calls would seem to be an unnatural presentation of a stimulus.

Chapter 3

Olfactory Repellents

Smokes and fumes cause living organism to move away and used for the trapping/killing of rodents from their burrows since antiquity. Apart from this, odour associated with their predator may form an aversion to mammal pests, thinking the presence of predator, making them to flee. Olfactory cues may serve as conditional stimuli to which learned aversions can be formed when paired in the presence of toxicants or irritants (Waldvogel, 1989; Clark and Smeraski, 1990). The most effective animal repellents will likely be those that produce condition aversions (*i.e.*, avoidance rather than escape behaviour) in the target species (Rogers, 1974; Mason and Clark, 2000; Werner *et al.*, 2008).

3.1. Smokes and Fumes

It is common to cough for a few minutes after breathing in smoke or fumes from a fire. Smoking and capturing animals is useful in areas which would be difficult to access with standard traps *viz.*, hollow trees, logs or rock jumbles. Holes can be marked and nets fixed securely over them (mist nets are appropriate, fishing nets are generally too coarse or of materials which can cut and injure a struggling animal). Then fire is lit on the base of the primary hole and covered with green leaves causing a thick smoke which drive out sheltering animals to be caught in nets (Barnett and Dutton, 1995). Traditionally people use smoke to drive out rodents from the holes to trap and kill or to allow predators like dogs and cats to capture. Burrowing rodents can also be killed using smoke grenades or engine exhaust piped into the sealed burrow (DeMoranville *et al.*, 2000). The common method of using a flexible hose to pipe automobile or tractor exhaust fumes into the burrow is perhaps most effective because it forces gas through all open burrows almost instantly (Cummings, 1962). Fumigation, a successful method for controlling some rodents, is of limited effectiveness in case of gophers. Gopher burrow systems are extensive and portions of them are blocked off by earth plugs as the gopher occupies various portions at a time and gophers quickly seal off their burrow when they detect smoke or gas. Burning tyres and using old engine oil or creosote are reported to be repellents for bush pigs, *Potamochoerus porcus* to protect maize crop at cob stage (Grange, 1986).

Smoke has been used to disperse birds from nesting and roosting sites. Smoke pots were used to deter starling roosts and the birds were found not leaving the roost on the night of treatment, but not returned on subsequent nights (Seubert, 1964). Smokes were used to deter Bobolinks, *Dolichonyx oryzivorus*, which is a serious pest of rice crop in Bolivia but with very little satisfaction (Renfrew and Saavedra, 2007).

Smokes of acids and similar substances are corrosive and cause irritation on exposure and dangerous to inhale. Ammonia odour repels many animals with a few exceptions like naked mole-rats, *Heterocephalus glaber* which lack neuropeptides associated with the signaling of chemical irritants and thus adapted to live in a challenging subterranean environment with extremely high levels of ammonia and carbon-di-oxide (LaVinka *et al.*, 2009). Gassing is an expensive method of control and ordinarily is not recommended for animals or birds (Cummings, 1962). It is difficult to maintain lethal concentrations of gas, particularly if the soil is not moist and tight and unless the gas is applied under pressure.

3.2. Aerosols

The principle behind the use of avian aerosol repellents is the same as that exploited in the use of ortho-chlorobenzylidene malononitrile (CS) and chloroacetophenone (CN) tear gases for human crowd control (Yih, 1995; Anderson *et al.*, 1996). A fog is formed from a mixture, comprising water and a chemical bird/ animal aversive agent, preferably the mixture also includes a dispersant. The water fog is formed by heating and or pressurizing so that it vapourizes and directed into the atmosphere towards the animal to be repelled. Some of the chemical agents used for aerosol fogging are methyl or dimethyl anthranilate, esters of phenyl acetic acid, ortho amino benzoic acid derivatives, cinnamic acid derivatives etc. Aerosol delivery of chemical repellents might work to effectively target birds in the air prior to landing. The nociceptive system that mediates the detection of orally presented irritants also innervates the mucosa of the nose and eyes.

Thus tear gas is principally used to elicit a change in behaviour, *i.e*, to promote avoidance of an area based on severe sensory irritation (Stevens *et al.*, 1998). Aerosols especially of methyl anthranilate acts as a sensory repellent by irritating the bird's taste buds, skin and trigeminal chemoreceptors in the beaks, gizzards, eyes and mucous membranes. Aerosol delivery strategies have also been used in agricultural contexts to effectively disseminate insect pheromones in communication disruption programs. It was found that 'puffer can' (aerosol-releasing devices) provide an efficient means to target insect pheromone receptors under field conditions (Shorey *et al.*, 1996). In order to determine the efficacy of such a deterrent strategy for birds and the nature of the behavioural response to aerosols, laboratory trials were conducted in which European starlings, *S. vulgaris* were exposed to short (30 sec) aerosol bursts of methyl anthranilate.

Results illustrate that birds demonstrate a clear irritation response to the aerosol with no evidence of habituation (*i.e.*, reduced responsiveness) under repeated exposures.

Methyl anthranilate is a potent avian chemosensory irritant which repels the birds than killing it. Stevens and Clark (1998) has explored the possibility of employing methyl anthranilate aerosols as a bird deterrent strategy. The behavioural response of starlings, *S. vulgaris* were studied with each of three aerosols: water or yucca extract (as control) and ReJeX-iT TP-40™ (40% methyl anthranilate) and found that starlings were irritated by exposure to the methyl anthranilate aerosol. Moreover, starlings did not habituate to repeated exposure to methyl anthranilate aerosols. The threshold of irritation for starlings to formulated aerosol was found as 8% methyl anthranilate. The number of molecules impacting the eyes of starlings when aerosol sprayed at concentration of 8% methyl anthranilate solution was estimated as 3.4×10^{13}, based on average particle volume and integration over a 10 sec exposure (Stevens and Clark, 1998). In the laboratory, concentrations of 0.5 to 2.0% were reported as effective repulsive concentrations in feeding trials, while 0.1 to 1.0% effectively repelled birds from consumption of treated water during drinking trials (Mason *et al.*, 1989; Clark *et al.*, 1991). Methyl anthranilate aerosols were used in aerodromes to disperse birds like migratory swallows and killdeers, *Charadrius vociferus* (Engeman *et al.*, 2002). Aerosols provide a practical and efficient solution to traditional bird hazing problems and merit further investigation and refinement as an avian deterrent strategy. A possible application of aerosol repellents strategy in the context of roost dispersion is also to be tested and proved.

Clark and Shah (1992) have applied this technology to predict olfactory-mediated foraging behavior in Leach's storm petrels, *Oceanodroma leucophrys*. Application of aerosol plume models to the planning of bird hazing operations will allow system managers to optimize the placement of aerosol sprayers in order to maximize the likelihood of targeting birds in flight with an effective dose. Computer simulations of plume behaviour must incorporate data on prevailing wind conditions, bird flight patterns over the protected area, estimates of aerosol sprayer coverage and avian detection thresholds. Initially this necessitates intensive field observations, but avoids inefficiencies and errors in the setting of hazing devices within the protected area.

Further, smaller droplets in the 10μ range are 27 times more effective than 30μ droplets and 5μ droplets even 216 times more effective for the same quantity of product applied. It has been shown that droplets below 10μ are inhaled by birds much more and therefore show an increase in effectiveness. Large droplets not only are relatively ineffective, they also tend to drop to the ground prematurely and are lost. Smaller droplets stay in the air much longer and disappear by evaporation and do not drop to the ground (Vogt, 1997). The generation of an invisible fog at unexpected intervals to make the area totally unacceptable to the birds until they leave forever is needed. Capsaicin aerosols are also used and found effective against bears (Smith, 1984) and elephants (Osborn, 2002).

Capsicum aerosol spray has reported as an elephant repellent over short (20-30 m) and intermediate (50-100 m) ranges (Osborn and Rasmussen, 1995). Aerosols are usually effective repellents for short durations and weather parameters and local topography is also to be considered before using aerosols.

3.3. Predator Odour and Others

Animals perceive odour very effectively, which is evident from the trained dogs used by police departments and 'search-and rescue squads' to track both suspects and victims through olfactory cues. Use of this odour perception of animal as repellent is a relatively new approach of pest management. Now, there are a number of odour repellents marketed to repel deer, rabbits and other mammals from browsing on vegetation (Hygnstrom *et al.*, 1994). Some of these products might be appropriate for short-term protection of valuable landscaping plants and fruit trees (Cleary and Dolbeer, 2005). These include products that are applied directly to the vegetation and general area repellents (e.g., predator urine). These olfactory cues also serve as conditional stimuli through which learned aversions can be formed. Pest birds and animals that perceive the odour of the potential predator may flee from a restricted area thinking of the predator.

Recognition and avoidance of odours of sympatric predators by mammalian prey has been well documented (Fulk, 1972; Stoddart, 1980). The use of predator odour has reportedly reduced damage caused by several species of mammals, including rat, *Rattus* sp. (Vernet-Maury, 1980), snowshoe hares, *Lepus americanus* (Sullivan and Crump, 1984, 1986; Sullivan *et al.*, 1985a; Sullivan, 1986), voles, *Microtus* sp. (Sullivan *et al.*, 1988a), pocket gophers, *Thomomys talpoides* (Sullivan *et al.*, 1988b), mountain beavers, *Aplodontia* sp. (Nolte *et al.*, 1994a), wood chucks, *Marmota monax* (Swihart, 1991), mule deer, *Odocoileus hemionus* (Sullivan *et al.*, 1985b; Melchiors and Leslie, 1985) and white-tailed deer, *O. virginianus* (Swihart *et al.*, 1991). Herbivores have shown a general aversion to the volatile constituents in their predator urine (Sullivan, 1986; Sullivan and Crump, 1986). Predator urine kept in small tubes tied on trees of Japanese yews, *Taxus cuspidata* was tested against white-tailed deer browsing. Bobcat urine has significantly reduced the browsing of deer followed by coyote urine whereas urine of human or rabbit and water did not have any effect. Browsing was severe in control treatments even up to 80% damage in 14 days.

The repellency of bobcat and coyote urine is enhanced further by spraying it on the trees. Spray trials revealed coyote urine as more effective than bobcat urine in a trial period of eight weeks (Swihart *et al.*, 1991). Mule deer were repelled more by coyote urine than bobcat urine (Sullivan *et al.*, 1985b). Red fox, *Vulpes vulpes* urine and mustelid anal gland secretions were found to repel snowshoe hares (Sullivan and Crump, 1986). Gorman (1984) found that voles, *Microtus arvalis* and *M. agrestis* responded aversively to anal gland secretion of a stoat, *Mustela erminea* but not of a non predator, guinea pig (*C. porcellus*).

Predator odours including a synthetic fox urine mixture were reported to significantly reduce vole, *Microtus montanus* and *M. pennsylvanicus* attack on apple trees (Sullivan *et al.*, 1988a). Bobcat, *Lynx rufus* urine and the predator odour compound, 3-Propyl-1,2-dithiolane and 2-Propylthietane are reported to be effective against porcupines and prevented them from damaging ponderosa trees (Witmer and Pipas, 1998). Dichloromethane extract of bobcat urine was found to repel white-tailed deer from browsing (Swihart *et al.*, 1991). Sulphur free coyote urine was found less offensive to mountain beavers, *Aplodontia rufa*. Thus the sulphur compounds in coyote urine appear to be needed for repellency (Nolte *et al.*, 1994a). Bobcat feces, chicken eggs and coyote, *Canis latrans* urine were found as deer repellents (ElHani and Conover, 1995). The synthetic deer repellent, Felinine® is a non-volatile amino acid found in the urine of domestic cats and bobcats (Westall, 1953; Mattina *et al.*, 1991) as well as the acidic component of lion, *Panthera leo* feces (Swihart, 1991). Andelt *et al.* (1991) tested coyote urine on captive mule deer and found effective when alternate foods were available.

He further found even 100% coyote urine and did not completely suppress browsing by hungry elk but was deterred from feeding on alfalfa (Andelt *et al.*, 1992). Mink and coyote urine deterred mountain beavers from damaging Douglas fir trees (Nolte *et al.*, 1993a, 1993b). Belant *et al.* (1998b) found that predator (bobcat and coyote) urine kept in scent darts near the fields has decreased white-tailed deer visitation to feeding station by 15 to 24%, compared to untreated control for two weeks thus concluded as slightly effective as a chemical barrier. North American deer have also been shown to avoid predator urine (Melchiors and Leslie, 1985; Muller-Schwarze, 1972, Sullivan *et al.*, 1985b; Swihart *et al.*, 1991). Although direct application of predator urines to food can suppress feeding by deer (Sullivan *et al.*, 1988a; Swihart *et al.*, 1991), predator urines were only marginally effective in excluding a high-density population of white-tailed deer from establishing feeding areas and were ineffective in reducing deer use of trails.

In feeding station experiments, coyote hair (17 g) placed in bags adjacent to or in front of a trough of whole kernel corn acted as repellent to white-tailed deer and intrusions by deer at treated sites decreased by 48 to 96% and corn consumption by 59 to 91%. In sites with coyote hair placed adjacent to trough or little away, the repellency decreased after 3 weeks. In the deer trail test, use of trails did not differ between the pre-treatment and treatment periods for the control or treated trails. Coyote hair therefore served as an effective repellent to keep deer from a desired food source and should have utility in protecting limited, discrete sites. However, coyote hair did not deter deer from moving along established trails (Seamans *et al.*, 2002).

Dried blood, predator animal scents, old shoes and a myriad of other concoctions applied to rags or other materials and exposed in trees as area repellents have met with little success. A mixture of fresh cow dung and lime has proved an effective repellent for small antelopes (Hey, 1964). Volatile compounds (amines and volatile fatty acids) of fermented egg are regarded as an attractant to coyotes and repellent to deer and porcupines (Bullard *et al.*, 1978; Witmer and Pipas, 1998). Chicken eggs was tested on captive mule deer and found repulsive (Andelt *et al.*, 1991). In addition, many naturally occurring plant secondary compounds are known to deter feeding in herbivores, by repulsive smell. Witmer and Pipas (1998) stated that spear mint oil (17%) capsaicin (0.25%) and putrefied egg solids (36%) are very effective in pen trials against porcupines whereas garlic oil spray at 10% as not effective. European badgers, *Meles meles* can be repelled from gardens and lawns with repellents such as an absorbent rope soaked in substances like old diesel oil or renardine (Neal, 1986).

Gorman (1986) tested African elephant temporal gland secretion as an elephant repellent with somewhat ambiguous results. Male African elephants secrete different hormones from their temporal glands when they are in and out of musth (Rasmussen *et al.* 1996). Synthesized gland secretions from periods of intense fear (*i.e.* culling) could be used as a repellent, if animals produce a fear pheromone (Osborn and Anstey, 2002). The urine, feces and glandular secretions released by a badger are unbeatable attractant to other badgers. So once a badger is caught in a trap, traps are used continuously if there are other pest badgers in the area to attract them and trap them easily. The combination of disturbed soil mixed with badger smell act as olfactory attractant for the next badger (Minta and Marsh, 1988).

Chapter 4

Tactile Repellents

Tactile repellents make a tacky feeling on the animal and if strong enough, it makes them to move to other place of comfort. Tactile seed coating (clay based) or perch repellents (polybutene products) are reported effective for some pest bird. Modification of the perch or perching place also plays an important part in repelling birds from roosting and nesting or in other words using the area.

4.1. Tactile Seed Coating

An alternative to lethal techniques and to chemical repellents is a bird-resistant, nontoxic seed coating. This method has proven effective in cage and small enclosure trials and warrants large field trials. In cage studies there was almost a 6 fold difference between feeding rates with coated and uncoated seeds (Daneke and Decker, 1988). The coated rice becomes sticky when wet and fouls the birds' bills so that they cannot feed efficiently. As a result, they switch to more easily eaten alternate foods. Clay coatings also accumulate soil and detritus that provide camouflage from birds (Daneke and Decker, 1988). Cowbirds avoid seed coated with agricultural lime (5% wt/wt) when the pH exceeds 12.3. A second underlying component mediating repellency is based on avoidance of particulates. If the particulate seed coating consists of particles sized ~63 to 150 μm, and has a pH of 11.4 or less, the repellent potency is about half that observed for raw unprocessed lime (Clark and Belant, 1998).

Clay-based seed coatings that become tacky when wet are effective bird repellents under some conditions (Avery *et al.*, 1989; Decker *et al.*, 1990). In clay coating treatment in rice seeds, the overall loss in control plots was estimated to be 36.5% of their sprouts compared to 17.0% in the treated plots. Observations showed that the treated plot received roughly 14 times less bird use than did the control. Average feeding rates of red-winged blackbirds, *A. phoeniceus* was 1.5 and 8.4 seeds/min in the treated and control plots, respectively. The results of this study are consistent with laboratory findings and with predictions from foraging theory (Decker and Avery, 1990). The birds also concentrated their feeding activity in the wet areas of the field where drainage was incomplete and where seeds were easier to remove from the mud. The activity recorded in the treated plot indicates the birds learned quickly to avoid the coated rice and to use the untreated plot instead. The birds required considerably longer time to handle and eat a coated seed than an uncoated one. The field observations are consistent with the original premise that if a normally preferred food is made sufficiently difficult to locate or to process, then birds will move to forage elsewhere on food that is easier to handle. Clay-coated seeds that become sticky when wet have been found to be repellent to captive red-winged blackbirds (Daneke and Decker, 1988). However, when bird numbers are high and when alternative foods are relatively unpalatable or sparse, clay-based coatings confer little protection.

In an aviary test conducted to study the effectiveness of dolomitic lime, activated charcoal, Nutralite® (a silica-based compound) and white quartz sand as feeding repellents for brown-headed cowbirds (*M. ater*), consumption of treated millet (1% to 4% g/g) was less than that of untreated for all particulates except Nutralite® at 1% g/g. Greatest reductions in consumption occurred with lime-treated millet, followed by charcoal, Nutralite® and sand (Belant *et al.*, 1997c). Further studies states dolomitic hydrated lime mixed with millet or whole-kernel corn at 25, 12.5, and 6.25% (g/g) reduced cowbird and Canada geese, *B. canadensis* feeding in cage trials. Reductions in total food intake and body mass occurred for both species during no choice tests with millet or corn treated with lime (25% g/g) (Belant *et al.*, 1997d). Nutralite® applied to turf at 2,568 kg/ha reduced overall goose presence on treated plots for 3 days but suppressed goose grazing for 1 day only. It was also reported that application of lime to enclosed 10x10 m grass plots in powder or slurry form at an application rate of 544 kg/ha also reduced goose feeding on treated plots for 2-3 days (Belant *et al.*, 1997b). But lime applied to turf at 270 kg/ha did not suppress grazing by Canada geese (Belant *et al.*, 1997a). Study with free ranging white-tailed deer, *O. virginianus* revealed a low consumption of corn treated (4% g/g) with dolomitic lime or activated charcoal than corn treated with Nutralite® (silica) or quartz sand. Corn treated with sand did not reduce consumption by deer relative to untreated corn whereas lime was highly effective (Belant *et al.*, 1997c).

Coal tar and other derivatives are used as seed treatments for repelling or causing aversion to birds and protect the seed from feeding (Schafer, 1978). Seed coating with Portland cement or plaster in rice acts as a tactile repellent for birds especially of red-winged blackbirds, *A. phoeniceus* but no mortality occurred when birds were offered only treated rice over a 4 day period (Dolbeer and Ickes, 1994). Though these seed coatings are found to create an aversive reaction in the pests, it should be carried along with other management strategies to have an effective control.

4.2. Tactile Roosting Repellents

Tactile repellents involve the use of sticking substances that discourage birds because of their 'tacky' feel. When these tacky materials get on the birds feet and is unpleasant for them to roost, they quickly learn to avoid the areas where the material has been applied thus moving to other areas of less concern. Polybutenes and polyisobutenes are used as tacky roost or ledge repellents often to repel pest birds from buildings and structures and are usually non toxic (Schafer, 1978). Such repellents are designed to be just sticky enough to discourage landing but not sticky enough to trap the bird. These repellents are formulated as liquids or gels or as sticker strips and occasionally applied on some specific areas of farm or trees to prevent roosting. Polybutene products of both tacky pastes and liquids repel birds from ledges or other roosting structures (Tim, 1983). A plastic gel called 'Scarecrow Strip' is used on buildings to deter starlings and feral pigeons from roosting (Seubert, 1964). Soft, sticky repellents are non-toxic materials used to discourage starlings from roosting on ledges or roof beams, for example, Roost-No-More®, Bird Tanglefoot®, 4-The-Birds® and others. These polybutene tactile repellents are reported to effectively discourage birds such as pigeons, starlings, house sparrows etc., from perching on ledges, beams or similar areas. Tactile repellents to deter perching contain polybutene and may contain other substances to induce a chemical reaction that gives the bird a mild 'hot foot' (Bishop *et al.*, 2003).

These products often contain other ingredients including mineral oil, lithium sterate soap, diphenylamine, zinc oxide and castor oil (Tim, 1983). Mineral oils are used as bird repellents by making the place unattractive for roosting or loafing of pest birds thus act as tactile repellents (Schafer, 1978). Clark (1997) reported that starlings became agitated and hyperactive after their feet were immersed in 5% oil extracts of the spices cumin, rosemary and thyme. Starlings avoided perches treated with either R-limonene, S-limonene, limonene and methiocarb. Hollow metal perches containing a wick treated with the toxicant fenthion were used previously to control pigeons, house sparrows and starlings in and around buildings (Cleary and Dolbeer, 2005). In theory, these repellents cause irritation to the bird through contact with the dermis on the foot and birds avoid such areas subsequently.

Often a masking tape is pasted on the surface needing protection and then the repellent is applied onto the tape. This increases effectiveness on porous surfaces and makes removal, if desired, easier. Over time, these materials lose their effectiveness and must be replaced (Johnson and Glahn, 1998). While effective, polybutane based repellents are thermally labile and melting can deface structures to which it is applied (Mason and Clark, 1995). These polybutenes are reasonably effective for periods of 1 year or more against birds especially nesting or roosting sparrows. They are messy and should be placed on tape or sealed masonry surfaces so they can be removed. They lose their tackiness after they become hardened by changing weather or covered by dust (Fitzwater, 1994). These repellent materials are negatively affected by temperature extremes and collect dust or debris on their surfaces. Therefore if the bird problem persists, reapplication is required after 6-10 months. Dusty environments can substantially reduce the life expectancy. Once the material loses effectiveness, it is necessary to remove the old material and apply a fresh coat (Cleary and Dolbeer, 2005). To be effective, all perching surfaces in a problem area must be treated or the birds will move a short distance to an untreated surface. These repellents are unpopular because they are not cost effective and also difficult to apply.

Another approach of repelling is using perches coated with glue to frighten birds from a farmer's field. Glue applied perches are also used to deter birds from rice fields

(Reidinger and Libay, 1979). The authors found the glue to be effective during the short treatment period (5 to 8 days). A viscous product named 'Scaraweb' is placed on gooseberry bushes with some success against many pest birds. This method would take advantage of some birds which alight on tall weeds or sticks within the field to feed on. If a few of the birds adhered to the perches, the glued birds emit distress calls and the remaining birds would learn to avoid the treated fields. Reidinger and Libay (1979) used a bird glue by mixing the saps of two species of trees, *Artocarpus* spp. and boiling with low heat over an open fire until very sticky. A total of 120 perches were coated with glue and placed in rice fields. It was reported that glue perches were very effective in reducing the bird population in rice fields. In three days, 9 tricoloured munias (*Lonchura malacca*), 4 white-bellied munias (*L. leucogaster*) and 18 other birds representing three common rice field species became stuck in the glue and emitted distress calls. *Lonchura* quickly learned to avoid plots that had the perches and continued to avoid these plots for at least 5 days after the glue had lost its adhesive qualities (3 days). A reduction of 85.9% in granivorous bird species was reported in treated plots with respect to reference plots for one to four days.

Surface or Perch Modifications

Tubular steel beams are much less attractive as perching sites for starlings and pigeons than are L-beams. A perch size of 3/8" x 27" is recommended for sparrows and starlings and 1' x 24" for pigeons. For pigeons, a flat surface (non-grasping) is well suited for sitting (Jackson, 1978). Cliff swallows, *Petrochelidon sp.* demonstrate preferences for rough and unpainted surfaces for nesting (Brown and Brown, 1995). Gorenzel and Salmon (1982) and Salmon and Gorenzel (2005) described methods to modify surfaces to deter swallows from nesting on them. Such methods include surface modifications such as anti-perching spines, smooth strips mounted at an angle of at least 45°, panels of glass, sheet metal or paint to create a surface unfavourable for cliff swallow nesting. Porcupine wires (Nixolite® and Cat Claw®) are a permanent type of mechanical repellent. They are made of stainless steel or plastic prongs with various types of sharp points extending outward at all angles. The sharp pointed wires inflict temporary discomfort and cause birds to avoid landing on these surfaces. High density polyethylene sheets were attached to bridges using butyl based sealant tapes to make the surface unattractive for cliff swallow to perch or make nests. This act as a smooth surface whereas these swallows prefer rough surfaces to build their nest and this treatment was found to reduce 69.1% of nesting in the bridges (Conklin *et al.*, 2009). These devices or materials are used to prevent landing or roosting of birds in a particular site. Some birds tend to roost on wires or in trees and thus tactile repellents offer little use in the management of birds like starlings.

Chapter 5

Gustatory Repellents

Most repellents work by triggering a primary, innate response in the target species to the compound's distasteful or unpalatable taste. Taste aversion is one way of repelling the pest from its favoured food. But the taste of animals may vary profoundly from species to species. Some repellents are also capable of inducing post ingestional malaise that the consumer subconsciously associates with taste of the treated food, thus establishing a conditioned aversion to the treated food. If this conditional aversion is created due to gustatory repellents, it will be of much useful than frightening them after eating.

5.1. Taste Aversive Chemicals

Many plants contain bitter components, because of this taste characteristic and because of the bitter compounds are often toxic, such substances and the plants that contain them are regarded as generally unpalatable to wildlife (Garcia and Hankins, 1975). Taste repellents can be divided into primary and secondary repellents. Primary repellents are agents which upon first exposure causes irritation resulting in avoidance. Secondary repellents are not immediately offensive but cause illness or an unpleasant experience following ingestion so that the animal avoids the treated food in future. Conditioned food aversion is due to negative gastrointestinal consequences (e.g., nausea) after the ingestion and the integration of sensory (flavour) and negative post-ingestive consequences (*i.e.*, effects of nutrients or toxins on chemo, osmo or mechano-receptors in the gut and brain) (Provenza, 1996; Wang and Provenza, 1996). The sensitivity of taste receptor in birds is similar to that of mammals and is species specific in their response to various chemicals (Moore and Elliott, 1946; Duncan, 1960; Berkhoudt, 1985; Ganchrow and Ganchrow, 1985; Mastrota and Mench, 1995). Gustatory receptors are located in taste buds located throughout the oral cavity of mammals and birds (Berkhoudt, 1985; Ganchrow and Ganchrow, 1985) and require a more intimate contact between the source of the chemical signal and the receptors (Mason and Clark, 2000).

Different programs on evaluating compounds for bird repellency had been done for many years with more than 200 test chemicals and found that odour plays a little part in repelling or deterring birds whereas taste is more important (Welch, 1967). Several food treatments have been tested for their effectiveness for repelling birds and mammals. D-pulegone and mangone found naturally in certain plants has been demonstrated to deter blackbirds, starlings, northern bobwhites (*Colinus virginianus*) and domestic dogs from feeding (Mason *et al.*, 1989; Mason, 1990; Mastrota and Mench, 1995; Avery *et al.*, 1996; Mason and Primus, 1996; Wager-Page and Mason, 1996). Belant *et al.* (1997b) conducted tests to compare the relative repellency of these chemicals on caged brown-headed cowbirds, *M. ater* and concluded that mangone is less effective than d-pulegone and would likely be ineffective as a repellent for seed treatment. However, d-pulegone was found to deter feeding of brown-headed cowbirds in choice and no choice test at 0.1% (g/g) millet bait. Tannins are the best known chemical components associated with bird resistance in agricultural crops. The tannin in such crop varieties gives an astringent taste to birds thereby inducing avoidance.

Many animals and birds are reported to avoid plants high in tannins and do not thrive when fed on high tannin diets (Rogler *et al.*, 1985). Cinnamamide gives an unpleasant taste in the mouth of animals and bird found to repel rock doves (Watkins *et al.*, 1995), Canada geese, house sparrows, greenfinches etc. (Gill *et al.*, 1998). Cinnamamide is also unpalatable to Norway rats and house mice (Crocker *et al.*, 1993; Gurney *et al.*, 1996). Cinnamic acid, a powerful inhibitor of trypsin was thought to be an aversive for pigeons but not, whereas cinnamamide exerted a strong aversion (Crocker, 1990). Guinea pigs, *C. porcellus* learn to feed on those parts of bittersweet nightshade, *Solanum dulcamara* that contain the lowest levels of toxicant (Jacobs and Labows, 1979). Other herbivores also exhibit this capability. For example, goats can learn to discriminate and avoid high tannin concentrations when foraging on blackbrush, *Coleogyne ramosissima*. These aversions are rapidly acquired within the first feeding bout (Provenza *et al.*, 1990). Guinea pigs avoid citric acid, avidly consume at least some carbohydrate sweeteners and show preferences for sodium chloride and sodium saccharin (Jacobs, 1978; Beauchamp and Mason, 1991). Sucrose octaacetate was found to deter feeding of cattle and guinea pigs (Jacobs and Labows, 1979). Quinine and sucrose octaacetate were reported to reduce feeding in guinea pigs but denatonium benzoate, denatonium saccharide, limonene, L-phenylalanine, naringin and quebracho did not (Nolte *et al.*, 1994b). Guinea pigs tolerate quebracho, a tannin compound but meadow voles, *M. pennsylvanicus* were reported to avoid it. Some rodents avoid denatonium saccharide (Davis *et al.*, 1986; Langley *et al.*, 1987), while others demonstrate a preference for it (Davis *et al.*, 1987). Even closely related organisms frequently show striking differences in their taste sensitivity to various compounds (Kare, 1971). Although the reasons for these differences remain obscure, it is plausible that they reflect differences among species in their evolutionary history or ecology (Freeland and Janzen, 1974; Lindroth, 1988).

Lithium chloride and cupric sulfate are regarded as emetic compounds used in honey baits as a technique for preventing black bear, *Ursus americanus*. Colvin (1976) reported the taste aversive behaviour of captive black bears with 20 to 80 g of lithium chloride in honey baits. Gilbert and Roy (1977) evaluated taste aversion as a method of preventing black bear damage in apiaries. They used 6g of LiCl in both brood and honeycomb baits. These baits were set inside beehives and kept on likely avenues of approach of bears to apiaries. A significant reduction of 53.48% in damage of bee hives per bear visit compared to apiaries left without baits is reported. Dorrance and Roy (1978) also reported fewer penetrations per bear visit through the fence inside the bee yards where 12 g of LiCl bait was placed when compared to unbaited yards. However, bear visits per yard and visits per bear were significantly greater at beeyards with 12 g of LiCl per bait than at unbaited yards. Thus, bears were more inclined to return to beeyards where bait was present and the presence of bait did not change the overall damage rate. So it is concluded that baiting with lithium chloride or cupric sulfate are ineffective in reducing bear damage in beeyards. Another emetine di-hydrochloride treated eggs when presented to raccoons (*Procyon lotor*), opposums (*Didelphis virginiana*) and striped skunks (*Mephitis mephitis*) in the field was found ineffective (Conover, 1990).

Advanced research on these taste components in animals resulted in the development of commercially available gustatory repellents. Curb® and Reta® are chemical repellents which are used on agricultural crops against both birds and mammals. Both of these materials are multiple ingredient formulations based on the ammonium, sodium and potassium salts of aluminum sulfate (alum). Alum is a bitter material that is distasteful to humans and many animals (Schafer, 1978). There appear to be instances of repellence, firstly of taste and then of taste and smell, being effective for a short period. When Curb® treated seeds were given to birds; they picked it up in their beaks, wiped each grain on the grass and then appeared to throw it right to the back of their throats, presumably to miss the taste buds on the tongue (Stone, 1979). Curb® sprayed on maturing barley effectively repelled sparrows for two days. Curb® sprayed on the buds of a forsythia bush to prevent damage by sparrows and finches are found to be very effective for three days, after which all the buds were stripped from the bush and left on the ground but not eaten. Bullfinches were repelled from the buds of cherries, damsons, plums, apples, pears, gooseberries, red and black currants during the winter months when treated with Curb® grade 10 @ 20 kg/acre. The same dose was found effective to repel woodpigeons, rooks and crows from a field of maize from the time of emergence to harvest. Curb® was used to repel deer from damaging rose for a longer time of almost 6 months. Dusting of Curb® was found effective to repel rats and made them to evacuate the treated buildings. Reta® sprayed on city garbage was found to repel starlings, lapwings, partridge and seagulls (Stone, 1979). Although all these formulations have shown some degree of repellency, effective application rates are often extremely high (Schafer, 1978). Another gustatory repellent, Sucrosan® is used against wild boar in the form of wheat and maize based food pellets with phosphorous acid as the active ingredient (pH 2). These pellets attract wild boars by its odour and once it has eaten the pellets, the phosphorous acid would unfold its flavour. This being a disagreeable experience for the animals would lead to a future avoidance of the area by a learning effect. But this was tested scientifically and found less effective; however the frequency of damage events has altered positively but insignificantly (Schlageter and Haag-Wackernagel, 2012). Repellents such as soap and hot pepper sprays are distasteful compounds but cannot be used in food crops against animal damage.

Though some success is obtained from gustatory repellents against bird and animals, these repellents are yet to be studied systematically and subjected to further improvement. Gustatory repellents repel the pest bird or animal after few samplings but if found very effective, this may be used to cause a conditional aversion towards certain plants.

5.2. Non-Preferred Plant Borders and Buffer Zones

A wildlife border is actually a combination of a travel lane and food patch for animals and of improvement of edge redesigned to solve the farmers' troublesome problem of maintaining profitable cropland adjacent to woody forests. Wildlife border is considered as the perfect example of putting wildlife, farming, erosion (control) and forestry together in a technique acceptable to all. The first is to create a zone of reduced attractiveness between the protected area and the surrounding crops (Thouless, 1994; Seidensticker, 1984). This involves clearing secondary forest on the boundary and creating some physical distance between the boundary and cultivation. An optimal buffer zone should contain unpalatable crops (such as sisal) grown adjacent to sub-optimal elephant habitat (Thouless, 1994). Plants with aggressive growths like *Lespedeza* sp. are reported to have considerable promise to be grown in the wildlife borders (Davison, 1941). Native and introduced plants with aggressive growths even in partial shade and in competition with tree roots, able to prevent soil erosion, reseed and continue permanently on the site or with nature of self perpetuation can be good for wildlife borders. Taylor (1982), on the other hand, suggests that a multi-use buffer zone concept in poor agricultural area where livestock is also restricted. A buffer zone is defined as a physically delineated area, either within or adjacent to a protected area. Animals would be hunted in this area to restrict their movements from forest areas. However, a non preferred plant bordering the preferred cultivable crop is a practical and useful measure to combat animal problem.

A few researches exist on animal preferences for particular crops, but there are a few crops that some animals appear not to eat. A 16 ft strip of rye sown around the edges of a barley field on a wildlife refuge reduced jackrabbit (*Lepus* sp.) damage to the barley (Lewis, 1946). Since the jackrabbits do not relish rye, they apparently did not enter far enough to discover the barley (Howard, 1967). Places where deer are a problem artichokes, tomatoes, squash, rhubarb or chives may be planted to help buffer the rest of the plants. Figs, pomegranates, persimmons, olives, date palms and prickly pears are also relatively unpalatable to deer. Some of the most common deer-susceptible fruit and nut trees are apples, apricots, plums, prunes, cherries, peaches, oranges, almonds and grapes. Beans, peas, carrots and strawberries rank especially high in deer preferences thus requires good management (Marsh, 1991). Growing non-preferred or less damaged crops by wild pigs such as oilseeds, sun flowers, late sugarcane variety or even combination of crops by growing preferred ones in rows or guarded by non-preferred crops in impacted areas reduces wild pig damage (Chauhan, 2011). An obvious way of reducing the attractiveness of cultivated areas is to plant crops on the edges of protected areas, such as tea or oilseed, which are not consumed by elephants. Crops that elephants dislike (e.g. chilli, tea, sisal, tobacco, timber) are planted around food crops to create a buffer (Bell, 1984). Buffer crops (e.g. chillies) act as an unpalatable barrier for elephants and have the added advantage of being available for use in fires to deter elephants and potential surplus production can be used as a cash crop (Hoare, 2001a). Chilli fences were found to be highly effective in preventing crop damage by elephants in Assam, India (Davies *et al.*, 2011). The idea of using tea plantation as natural fencing against elephants could act as a buffer zone by putting more distance between the forest and agricultural land. But this barrier of distasteful crops has to be planted for at least for 1 km width around tempting cultivable crops (Osborn and Anstey, 2002). Farmers of Uttarakhand state in Indian Himalayas were found using a land race of wheat, 'Tank' as non-preferred by the rhesus monkeys, *Macaca mulatta* because of their long (7.03 cm) and spreading awns. Therefore, to protect the main wheat crop from damage by monkeys, they grow 'Tank' as the border rows and high-yielding varieties in the centre. Thus, whenever monkeys attack the wheat field, they first eat the spikes of 'Tank' which they dislike; thereby leaving the wheat field (Gupta *et al.*, 2009).

In Bwindi and Mgahinga Gorilla National parks, planting of live fences of *Ceasalpinia decapetala* alongside the park boundary on community land is made to prevent raiding of big mammals from forest area (Fungo, 2011). For such destructive animals as elephants, once within a field, there is little that can be done to reduce the raiding caused. The alternative is to plant such barriers that deter them from reaching the gardens. Mauritius thorn, *C. decapetala* is reported to be planted intentionally in Africa to act as a natural barrier against animals. However, there is very little data to suggest that this barrier is effective against elephants and other animals. It is also not clear that the fence will deter elephants and there is little conclusive evidence that this fence will work against primates or bush pigs. Cactus and sisal have also been tried but little systematic research exists on the effectiveness of these plants (Osborn and Anstey, 2002). Elephants have been found to render these buffer crops ineffective by simply traversing them en route to their preferred food crop (Bell, 1984) and in some cases have even been observed eating sisal (Hoare, 1992). This is not to say that this idea has no merit, but simply that it does not seem to work in isolation. Care should be taken in selection of barrier crops because this may attract other group of animals for feeding or shelter. Also, it appears that in some areas farmers do not participate in a barrier-planting scheme due to a concern that it will restrict their access to the forest. Further, this is not feasible in the developing world where land is at a premium and there is no evidence that this type of buffer zone has any impact as a deterrent.

Section II
Management Methods

Chapter 6
Mechanical Scarers or Protectants

Mechanical exclusion is considered to be the best, full proof technology for damage reduction by pest animals and birds. Bird netting is practiced in high value crops and aquaculture facilities where it is economical to use. Fencing when made and maintained properly can exclude a wide range of animals. A wide variety of traps are available to capture problem causing creatures from a small mouse to a huge elephant.

6.1. Barbed Wires and Antiperching Devices

More expensive, but long lasting bird and animal deterrence is obtained by mechanical structures with sharp metal projections such as Nixalite® and Cat Claw®. These sharp metal projections prevent the birds from roosting comfortably in an area (Fitzwater, 1994). Spikes/ antiperching spines employ sharp and pointed objects to create an unattractive and potentially dangerous landing site which the birds avoid (Salmon and Gorenzel, 2005). The commercial product consists of a narrow stainless steel strip from which emanates an array of needle like wires approximately 3-4 inches long. Bird coil uses flexible wire in the form of spring or coils to create an unnatural moving surface, which the bird will not prefer to land (Gorenzel and Salmon, 2008).

Anti-perching devices, such as spikes can be installed on ledges, roof peaks, rafters, signs, posts and other roosting and perching areas to keep certain birds from using them (Cleary and Dolbeer, 2005). Six different antiperch devices (Monofilament web, bird spinner, bird spike, Agcone, Agspikes and the combination of all) were tested on five bird species *viz.*, brown-headed cowbirds (*Molothrus ater*), fish crows (*Corvus ossifragus*), great horned owls (*Bubo virginianus*), barred owls (*Strix varia*) and black vultures (*Coragyps atratus*). No single device proved totally successful for all species involved in tests but these are appear to be effective for large birds, but less so for smaller species.

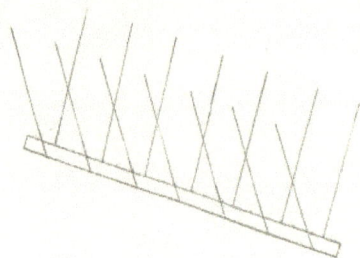

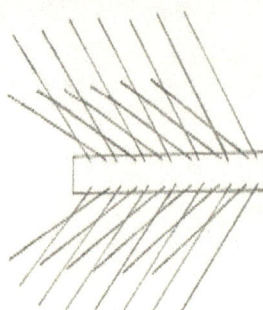

Simple wires   **Hedgehog wires**

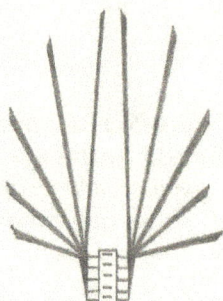

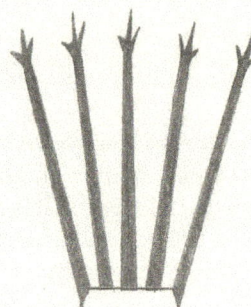

Needle wires   **Cat claw**

Figure. 6.1: Anti-perching devices

The most effective perching deterrent was a set of 17 stout spikes (AgSpikes) secured to the central portion of the unit that point up 0° to 30° from the vertical. These spikes were subsequently redesigned and combined with 9 metal bushings (3 for each arm of the sensor unit) that fit loosely on the arms and that were armed with 5 sharp spikes each. AgSpikes are found effective for all birds tested except owls, which were able to place their feet within the open spaces of the device and perch on the horizontal spikes. The combination device provided the best protection for all species; however, 100% deterrence was not achieved (Avery and Genchi, 2004). Categorically, larger birds such as owls and vultures require different devices than do smaller species [e.g., cowbirds and fish crows]. Seamans *et al.* (2007b) tested an antiperching device to deter brown-headed cowbirds, European starlings, red-winged blackbirds, rock pigeons and common grackles. A commercial antiperching device, Birdwire™ was tested in an aviary setting and reported effective in reducing perch use by all species. Blackbirds and starlings were however, capable of using the perches, but only for a short time.

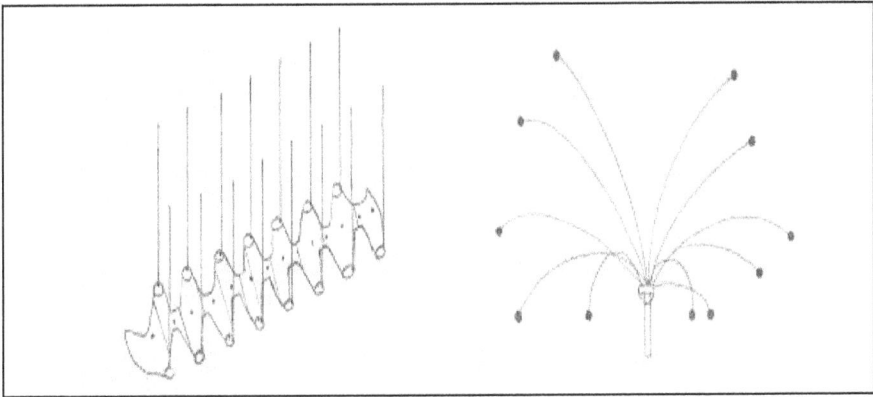

Figure 6.2: Anti-perching devices for specialized places

Barbed wire fences are used to discourage animal's entry in to agricultural lands. A barbed wire fence of three or more strands with 6 ft height can exclude many pest animals from entering agricultural lands. The wires are supported by temporary or permanent posts or trees. Live fence posts are formed by planting tree species like *Gliricidia sepium, Erythrina* sp., *Spondias* sp. or *Bursera simarouba* which aid in barbed wire fencing. In some situations, fences of barbed wire in combination with net are reported very effective (Wade, 1982). Low voltage electrified barbed wire fences are reported best for wading off animals with copious hairs as the points of the barbs penetrate to the skin and make an effective contact for the current (McAtee, 1939). However, it is better not to electrify barbed wires (Curtis *et al.*, 1994a) to prevent animal or bird tangle which causes deaths. These barbed wires are used to spin as a coil in the main tree trunks near the agricultural fields to prevent climbing and settling of monkeys in the trees while driving them out of the land with some means especially through human guards, as shown in photo 10.

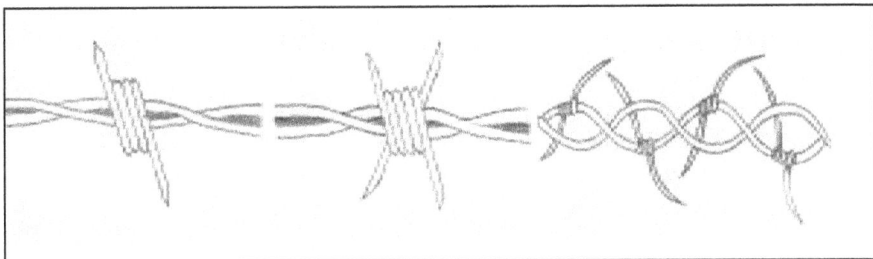

Figure 6.3: Different barbed wire models

6.2. Bagging

Preharvest bagging of fruits using cloth or polythene bags or even with the leaves is an age old practice to prevent animal and bird damage. Wraps or bags are sometimes used to protect individual or clusters of ripening fruits or vegetables. This is practical only on a small scale and mostly used for backyard gardens. Although this technique is most often used to protect fruits such as dates and figs from damage by birds (Popenoe, 1913; Chandler, 1958; Roach, 1985), it is sometimes used against mammals as well (Dowson, 1982; Kagoro-Rugunda, 2004). Apart from cloth and paper, other materials used for wraps/bags include matting of woven palm leaflets, cheesecloth, plastic and fiber netting and muslin (Chandler, 1958; Dowson, 1982).

Covering maize cobs by wrapping maize leaves around them reduced damage by rose-ringed parakeets, *Psittacula krameri* by 97% in the field and 82% in aviary experiments. Being hidden and camouflaged, the wrapped cobs may have escaped detection by birds and thus were not damaged. The wrapping treatment also provided some protection to uncovered cobs within the treated plot, possibly by reducing the density of visible cobs in this plot compared with the control. There was no adverse effect of the wrapping treatment on grain yield (Dhindsa *et al.*, 1992). Bagging of maize cobs with paper covers to prevent animal and bird damage is prevalent in India. Pre-harvest bagging of pear fruit, *Pyrus communis* with micro-perforated polyethylene bags of thickness 5µ, micro-perforations ca. 2 mm and size of 15 x 20 cm secured by stapling it tightly around the fruit peduncle at about 30 days after full bloom was reported to reduce the bird damage from 28.4% for non-bagged to nil for bagged fruit and found very effective (Amarante *et al.*, 2002). Bagging of fruits and squash prevents bird damage in the garden (Lin, 2003). The most common measures to avoid damage by primates include wrapping with cloth and guarding (Kagoro-Rugunda, 2004). Spillstoesser (1984) suggested covering ripening ears of sweet corn with paper bags to prevent raccoon damage, while sacking or cloth wraps are recommended for protecting dates and guava from damage by fruit bats or flying foxes (Ochse, 1931; Dowson, 1982). Cloth bagging of pomegranate fruits is a normal practice to prevent squirrel damage.

Preharvest treatments with covers and bags improve insect, disease and bird damage control apart from improving the quality of fruits (Berrill, 1956; Kehat *et al.*, 1969; vanDyk and Saayman, 1989; Pereira, 1990; Kitagawa *et al.*, 1992; Hofman *et al.*, 1997; Joyce *et al.*, 1997). For instance, brown bagging of apples is reported to stop codling moth damage (Bentley and Viveros, 1992). Bagging of fruits has many other advantages apart from preventing animal damages such as reduction in skin blemishes (Berrill, 1956; Joyce *et al.*, 1997; Tyas *et al.*, 1998), reduction in harvest and post harvest damages and reduces agrochemical residues in the fruit (Kitagawa *et al.*, 1992). Micro-perforated bags were reported to reduce skin blemish by 32.3% and increased the percent fruit accepted for export by from 27.2 to 63.2%. When sprayed with pesticides 3 days before harvesting, bagging reduced fruit residue of captan (non-systemic fungicide) and azinphos methyl (non-systemic insecticide) by 39.0 and 46.4%, respectively (Amarante *et al.*, 2002).

There are limitations on how effective this technique is against different birds and mammalian species. For instance, Dowson (1982) noted that in Israel hyenas, *Hyena hyena* are capable of tearing open wire-mesh bags to get at ripening dates. The point to consider in polythene bagging is that the material should be entirely perforated to avoid increase in humidity inside the bags, which promotes decaying. Damage by russet was observed in apples bagged with non-perforated polyethylene bags (Tukey, 1959). Further, bagging of all fruits is a labour intensive and time consuming process and thus can be practiced in high value crops only.

Photo 10: Coils of barbed wires in tree trunks to discourage climbing of animals

11: Netting to deter animal damage in corn

Photo 12: Bagging of corn to prevent bird damage

Photo 13: Netting in peach to prevent animal/ bird damage

Photo 14: Electric fencing to deter animal entry into agricultural land

Photo 15: Manual wading of monkeys from agricultural Land

6.3. Netting

Netting or screening can be the most effective method for protecting high value crops by completely excluding birds and animals from a site needing protection (Way, 1968). The technique is expensive, but costs may be justified in situations where other bird/animal control methods are ineffective (Fitzwater, 1978; Salmon and Conte, 1981). Although the use of properly mounted and maintained nets or acrylic fibres is expensive it can be compared favourably with the cost of employing people as bird scarers (Bruggers, 1982).

Earlier, plastic, fabric and wire nettings were used more often and the arrival of ultraviolet stabilized plastic netting in the early 1970s resulted in stronger and more durable nets which can be installed easily over large areas (Stucky, 1973). Physical barriers such as net or acrylic fibres are regarded as the most effective method of bird exclusion from agriculture fields. Xironet is a type of net made of a special blend of durable polypropylene used to cover crops or orchards especially vineyards. When kept in position it expands and gives a net hole size of 1 to 1.5 inch, which allows small insectivores birds to enter and does not affect plant. It is used in many crops like sunflowers, strawberries, currants, raspberries, apples and blueberries to give maximum protection against birds and mammals (Foster, 1979).

Nets were used for protecting blueberries (Hayne and Cardinell, 1949), sunflowers (Meylan, 1978), fruit crops (Stucky, 1973) and trees (Campbell *et al.*, 1981) against bird damages. Strawberries and similar fruits are covered with ordinary string netting to prevent bird damages. Viscose nets have been found effective and are being used more frequently for fruit trees, grapes and strawberries (Seubert, 1964). Meylan (1978) used nets for protecting sunflower crops from depredations by the greenfinch, *Carduelis chloris* and found useful except some incidences of bird trapped on nets and dead. Netting is reportedly used to prevent songbirds from feeding on high-value crops such as cherries, blueberries and grapes (Twedt, 1980).

Bird netting was found to be very effective in reducing fruit losses caused by frugivorous birds in blueberry, *Vaccinium corymbosum* with a tune of 75.6%, 74.7% and 88.9% with respect to control in three years of study (Vincent and Lareau, 1993). In vineyards, broadcasting of distress calls of starlings, finches and robins added with conventional methods like reflective tape, propane cannons and pyrotechnics significantly reduced damage when compared with conventional control (5.7% vs. 13.0%) whereas netting recorded the least damage (2.3%) (Berge *et al.*, 2007). Bird netting was set to be effective against wild turkeys, *Meleagris gallopavo* in vineyards but this technique is expensive and labour intensive to install, so it is not used at many vineyards (Coates *et al.*, 2010). Above all, an elevated frame work is needed to put nets in proper form without which the birds tend to perch on exterior on the net, eat or pick all the fruits within their reach (Simon, 2008).

Mist nets made of fine nylon strands woven into a net of various meshes (1 to 4 inches) can also be used to catch birds. Nets are usually seven feet high and eighteen to one-hundred feet long and when in use are suspended between poles. Strands of the net are very fine or are coloured to match the background. Birds fly into the nets and become entangled in the loose folds and can be trapped.

Overhead lines or wires strung in a grid or parallel pattern over the area from which the birds are to be excluded is also an effective technique. The spacing between the wires can vary from few inches to 25 ft or more and the efficiency decreases with increase in spacing. In wider spacing obviously there will be enough room for the birds to fly or settle between the wires. It has been suggested that the wires interfere with the bird's flight pattern, or that the thin hard-to-see line startles birds. Overhead lines do not physically exclude birds mostly instead they may represent a physiological barrier for some species of birds. Stainless steel wires are usually durable. The wires usually are strung parallel about 3 ft above the area to be protected, however heights above that are also found effective. The spacing between the lines depends on the species to be repelled. The following spacing have been reported effective: 20 ft for Canada goose, 11-13 ft for ducks, 2-20 ft for ring-billed gulls and 1-40 ft for herring gulls and most of them are used in water reservoirs (Gorenzel and Salmon, 2008). Overhead stainless steel wires at 16 m distance reportedly reduce roof-nesting by ring-billed gulls, *Larus delawarensis* and herring gulls, *L. argentatus* by 76% and 100% in first and by 99% and 100% in the second year (Belant and Ickes, 1996).

Netting is the only sure method of total exclusion. Netting ponds is the only mechanism by which the snowy egret (*Egretta thula*), green-backed heron (*Butorides striatus*), tricoloured heron (*E. tricolor*) and little blue heron (*E. caerulea*) can be deprived from fish ponds (Avery *et al.*, 1999). Fibers and netting can be used with success, but they are not practical for large cropped areas (Seubert, 1964). Netting systems give full protection of bird but it is costly and difficult to install and remove before the harvest. Cost effective tractor mounted net units for installing and removing from commercial vineyards and orchards are also available for easy and quick erection and removal (Fuller-Perrine and Tobin, 1993). The use of netting or screening for excluding birds depends on several factors, including the species to be excluded, size of the area needing protection, possible damage of the netting from severe weather and whether it will interfere with other operations at the site (Salmon and Conte, 1981). The most common netting made for bird control is made of polyethylene plastic, which is light weight and resistant to deterioration by sunlight. Nets with both small and large mesh size with variable weights are commonly available and used according to the need (Gorenzel and Salmon, 2008).

Nylon nets made as a wall using bamboo sticks are used to deter the entry of pest animals in agricultural lands photo 11. Nylon fish nets, 8 ft high, strung around forest plantations keep deer out during the season of damage (Mealey, 1969). Though it will not allow the easy entry of some animals, monkeys may jump over that or use them to climb and enter the protected area. The net must be regularly patrolled to release animals caught in it and to repair holes. It is a cheaper fence to build than a permanent 8 ft mesh wire one.

6.4. Fencing and Other Exclusions

The use of barrier fences to protect humans, domestic animals and some wildlife species from depredation began long before written history, perhaps with primitive man blocking the entrances of his caves for security from large carnivores. In various forms, barrier fences were developed and persist worldwide. Fitzwater (1972) reviewed the use of barrier fencing for more prosaic purposes in protection of crops and livestock and described numerous types, from earthen barriers to modern electric fences. Barriers of thorny shrubs, although ancient, are still commonly used for this purpose in many countries. The use of wire fences and later the barbed wire fences were found effective against various animals. Importantly, fence designs must consider the size, strength, intelligence and/or instinct and physical agility of the species to be repelled as well as the attraction of the crop or area for potential depredators.

Earthen barriers either in the form of breastworks, pits or trenches sometimes filled with water are commonly used by developing nations. A somewhat less expensive type used against bush pigs in South Africa consists of digging a series of holes, 2 ft deep and 3 ft in diameter. The excavated dirt is piled up in loose mounds between the holes (Thomas and Kolbe, 1942). Tribes in India build low ditches arraying the inner bank with several rows of sharpened bamboo spears to keep depredating antelope and nilgai from invading fields (Kumar *et al.*, 1963).

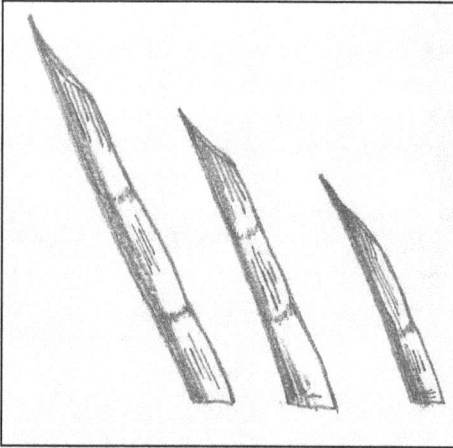

Figure 6.4: Sharpened bamboo fence

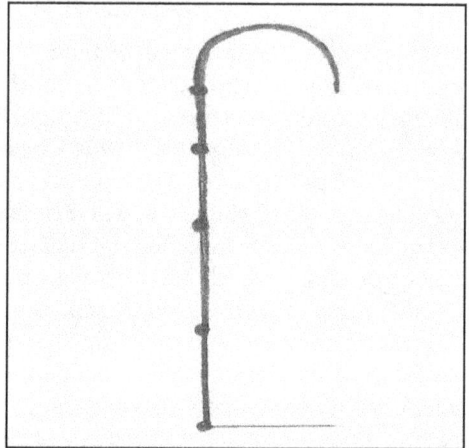

Figure 6.5: Fence design with a flip-flop top

Probably the oldest use is the piling up of thorny shrubs in more or less temporary barriers to prevent depredations. People use osage orange, black locust or multiflora rose, cacti and other thorny succulents as living hedges. Thorny twigs and branches of *Prosopis juliflora*, *Acacia nilotica*, *Ziziphus nummularia*, *Z. mauritiana* and some time naturally grown *Euphorbia caducifolia* and *E. tirucalli* are used as fence to protect agricultural crops from langur damage in India (Chhangani and Mohnot, 2004). Vegetative barriers in the form of logs and posts are also used. Brushwood fence used in some places is effective against cattle only but it rarely restricts nilgai and blackbuck and any form of fencing is of little use against them.

Thus the cost of providing protection to crops by barriers such as trenches, barbed wire and chain-link fences is prohibitive (Chauhan and Singh, 1990). Wire is probably the most common fencing material used world-wide today. It is flexible, light, easily erected and of reasonable cost. Wire can be used to keep out practically all animals except elephants and rhinos.

The last group used to build barriers includes a number of man-made materials-metal sheathing, burlap, plastic strips, nets, concrete and panels of plywood, asbestos, etc. Metal is commonly used in rat guards to prevent these agile rodents from climbing trees, ship hawsers, buildings, warehouses, fences, etc. Burlap has been draped over a 6 ft wire fence to contain African antelope that even though capable of clearing 8 ft fences would not jump the lower barrier because they could not see the other side (Fitzwater, 1972).

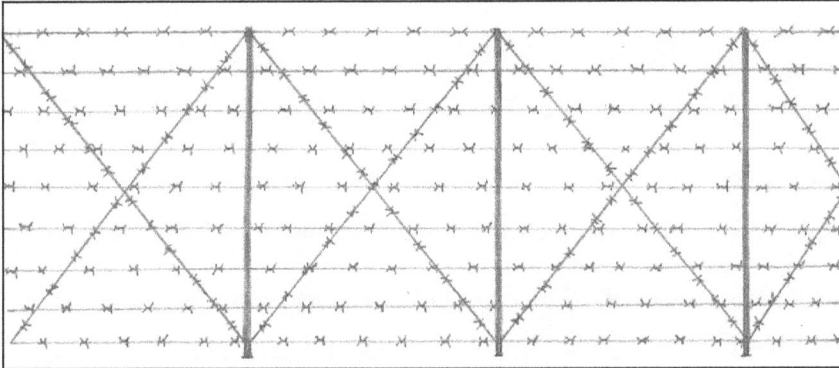

Figure 6.6: Barbed wire fence design

Exclusion is most often accomplished by the construction of rabbit-proof fences and gates around the area to be protected. Woven wire or poultry netting of a mesh not greater than 1 or 1.5 inches, 30 to 36 inches high, with the bottom 6 inches turned outward and buried at least 6 inches in the ground, should exclude all rabbits from the area to be protected (Johnson, 1964). One effective method of protecting crops from wild rabbit, *Oryctolagus cuniculus* grazing is to erect mesh fences, particularly where burrows are inaccessible and these fences are used for many years to protect crops from rabbits (McKnight, 1969). Enclosure trials conducted by McKillop *et al.* (1986) revealed hexagonal and rectangular meshes of 31 mm and 50 x 25 mm, respectively as the best to exclude all age classes of rabbits, *O. cuniculus* and 0.75 m as the most cost-effective fence height. However, Moseby and Read (2006) reported a 180 cm high wire netting fence with a foot apron and a curved 'floppy' overhang effectively contained most rabbits, feral cats and foxes during pen trials and proved effective with intensively monitored paddock-scale enclosures. Excluding rabbits from vulnerable crops can be done by wire-mesh fencing and when installed properly fencing can keep about 80% of rabbits out of protected fields (McKillop and Wilson, 1987; McKillop *et al.*, 1988). Sturdy sheep netting erected on timber and rail fences 16-18 inches at soil level and slightly below, keeps the badger from digging beneath (Neal, 1986).

Barrier fences were reported to perform well against deer even under intense deer pressure (Eadie, 1961; Caslick and Decker, 1977). A woven-wire fence 2.4 to 3.0 m tall is considered the most deer-proof design and can effectively exclude deer from larger areas and economically feasible for apple orchards (Caslick, 1980; Curtis *et al.*, 1994a). A specified gauge (11 to 14.5), high-tensile woven wire of tensile strength up to 121.5 kg/cm² with low stretch and high elasticity and breaking strength up to 816 kg can best suited for fencing for animals. The strong, elastic nature of the wire reduces stretch, sagging and damage when objects contact the fence. In addition, quality high-tensile wire receives Type III galvanizing, which can extend wire life up to 35 years in humid climates (Curtis *et al.*, 1994a). Fences typically installed to manage white-tailed deer damage include wire or plastic mesh fence (VerCauteren *et al.*, 2006). Rosenberry *et al.* (2001) successfully protected small plots (6 x 6 m and 12 x 12 m) with a 2.4 m tall plastic-mesh fence. Barriers are popular with communities in Kenya and found effective against crop pests such as zebras, *Equus burchelli* (Sitati *et al.*, 2005). Nonelectric fences for wild pig exclusion should be of net wire or diamond mesh construction with a min of 6 inch spacing (Littauer, 1993). Among the four different fence types tested against western grey kangaroos (*Macropus fuliginosus*), two of them, ringlock (netting) fence and fence topped by barb wire gave the greatest protection (Arnold *et al.*, 1989).

Figure 6.7: Ring lock fence design   **Figure 6.8: Diamond poultry mesh**

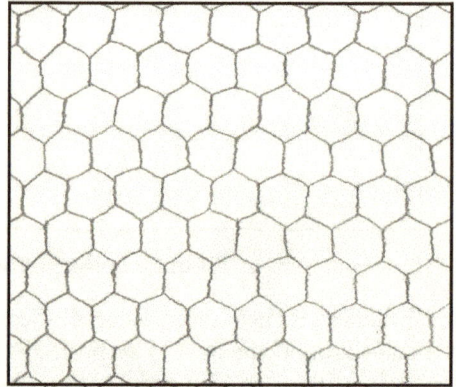

Fences do decrease the incidence of crop raiding elephants. In Negande, crop raiding incidents dropped by 65% after the erection of an elephant proof fence but this dropped to 42% the following season (Grant *et al.*, 2007). Whereas Sitati *et al.* (2005) reported passive barriers as largely ineffective, as elephants were easily able to break through them.

He further added that larger farms appeared to invest more heavily in improved barrier methods comprising wooden post and barbed wire fences at the expense of guarding effort. This reluctant guarding effort and barriers both increased the risk of crop raiding elephants in Kenya. The negative and additive effect of these barriers and complacent guards on elephant raids is explained as under. Elephants may have associated the presence of barriers with worthwhile rewards within and therefore targeted farms with such barriers.

More likely, however, is that farmers with such barriers, besides having fewer guards overall, were more complacent about actively guarding fields and so did not detect elephants soon enough to prevent them entering fields (Sitati *et al.*, 2005). However, a stout mechanical fencing of 2.4 m height with four cables, 6 straying wires and 3 barbed wires and with diamond mesh for 1.9 m height is found effective for elephants (Grant *et al.*, 2007).

Fencing appears to be most useful and cost-effective on small areas with intensive production which require full protection. It was reported that Hawaii Volcanoes

National Park removed goats (at a rate of 5,000/year) for a half century without any effect whatsoever upon the remaining goat population. After resorting to fences, the park zeroed out its goat population within a decade (Reeser, 1993). Fencing is regarded as a very good technique to prevent goat, pig, sheep and cattle populations entering into agricultural lands also. Concrete wall of about 4 to 5 ft height along certain strategic boundary areas or around at highly vulnerable crop fields to exclude wild boar will provide great relief to the farmers (Chauhan, 2011). Fencing should be considered for high value plantings that require year-round protection. For example orchards, landscape and tree nurseries, vineyards and other high value plants require perennial protection. Fencing can be a practical and cost-effective method for protecting small landscapes, gardens and small orchards. Cattle guards with wood frame and pipes appear to be an effective method of reducing animal crossings through fence openings (Belant et al., 1998c). Fencing is also used widely elsewhere to control wildlife movements as part of an attempt at control of foot and mouth disease (Taylor and Martin, 1987).

The adaption necessary to contain different species especially the size and strength physical ability of the animal, its intelligence and or instinct are to be considered for selecting a fence. An 8 ft chain link fence of 2 inch mesh will keep most animals out but not a pushy elephant or an inquisitive mouse. A rabbit being confronted with a tight mesh fence will generally try to burrow under it, so the fence must be sunk several inches into the ground to be effective. Pronghorn antelope are more inclined to crawl through a fence than jump over it even though they easily can, but a deer can clear over 7 ft. Norway rat burrows can go down several feet under, thus a 3 ft curtain wall is considered minimum (Fitzwater, 1972). To be successful in attractive and places nearby woodlands the fence must be closely maintained as well as tightly built. The last consideration is the attraction to breech a barrier and alternative food sources.

Fencing can be costly to erect and may require maintenance to remain effective, but it can be the most effective deterrent to animal damage. Gates, posts and hardware are additional materials that may be needed, which will add to the cost of fencing an area. Installation and maintenance costs should also be considered before deciding on fencing. Fence maintenance should include regular vegetation control along the fence. Thus, limitations on the use of exclusion fences include construction costs and the inability to exclude some predators or to remove them from fenced areas. Predators may gain access through damaged fences, malfunctioning electric fences or by jumping over or digging under. In circumstances, where predators tend to be contained within such enclosures, fences may serve to increase losses to predation rather than to reduce them (Wade, 1982).

Damage to wire fences by large animals is quite common. The mule deer, *Odocoileus hemionus* are noted in many areas as a major cause of fence damage through their habit of stretching and breaking openings in net wire mesh in order to pass through, thus providing ready access for other species. Armadillos, *Dasypus novemcinctus* are reported burrowing under net wire fences while the black bears are noted for their ability to defeat net wire by force (Wade, 1982). Foxes are reported to chew through a 0.9 mm gauge netting, as they did with plastic netting (Poole and McKillop, 2002), suggesting that thicker gauge netting is desirable. Phillippine rice-field rats, *Rattus rattus mindanensis* were reported climbing over the fence (Shumake et al., 1979).

Other exclusions:

Exclusionary methods and materials other than fences or full area enclosures have a long history of use in vertebrate pest control, particularly as a means of protecting young fruit and nut trees and tree seedlings for reforestation from deer (*Odocoileus* spp.), rabbits (*Lepus* spp., *Sylvilagus* spp.) and a variety of pest rodents (Marsh *et al.*, 1990). These exclusionary methods include various materials wrapped or tied directly on the tree trunks, the larger loose-fitting protective cylinders or other individual exclusives, shields or bands to prevent access to the upper tree portions via the trunk, mounding soil or other materials around the base of trees to restrict feeding or to make the habitat less favourable to pest species.

i. Tree trunk guards

Materials affixed directly on trunks: Today we have a variety of trunk protectors (wraps) that are specially designed, manufactured and sold for the purpose of protecting young orchard trees. They are manufactured of weatherproofed cardboard, plastic, aluminum foil, flexible aluminum mesh and other types of materials (Baer, 1980).

In earlier times, a variety of natural and discard materials were wrapped or tied around trunks of trees to protect them from bark-gnawing mammals. Natural materials of botanical origin that were locally plentiful and frequently used include cornstalks, dried twigs, bamboos, ropes of hay, thorny bushes etc. Empty cement, feed and fertilizer bags have also been used as trunk wraps. Many of these materials were inexpensive but not very durable; also occasionally provide harborage for insect pests and/or physical injury to trees when left on throughout the year. Natural materials such as cornstalks, bark strips, ropes of hay, etc., were effective to varying degrees, but they are time consuming to collect or prepare and labour-intensive to attach (Waugh, 1917).

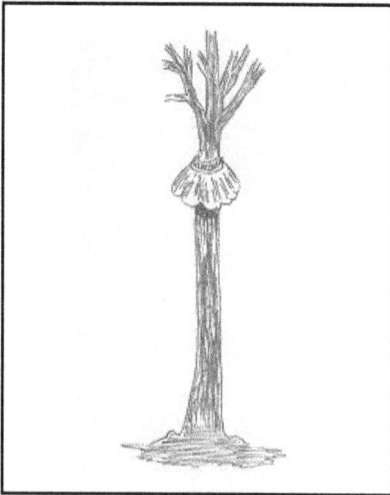

Figure 6.9: Tree tied with paper

Figure 6.10: Tree fitted with tin sheet

To prevent elephant damage to individual trees, a 13 mm mesh wire netting, wrapped around the trunk of mature tree stems prevented such trees from being extensively bark stripped (Gordon, 2003; Henley and Henley, 2007). But wire netting techniques did not protect trees from being uprooted or broken.

Cylinders for individual trees/seedlings: The use of individual mechanical protectors to guard the trunks of young trees or vines may also be considered a form of exclusion. Cylinders encircling trees/tree trunks form another category of exclusionary devices and are often used to protect young trees. Although these may be constructed from a variety of materials, cement cylinders, structures made using bricks and poultry netting are the most commonly used. Among the best of these mechanical contrivances are cylinders made from woven wire netting. Poultry netting of 1 inch mesh, 20 gauge galvanized wire 18 inches wide is commonly used against rabbits. This is cut into strips 12 to 18 inches wide and formed into cylinders around the trees. To afford adequate protection, these cylinders are braced away from the trunk to prevent rabbits from pressing them against the trees and gnawing through them (Johnson, 1964). Types of tree protectors commercially available include aluminium and nylon mesh wrapping and one made of treated jute cardboard. Plastic netting and ready-made tubes protecting forest tree seedlings from girdling, gnawing, clipping and/or browsing damage by animals. Plastic seedling protectors are effective against mountain beavers, *Aplodontia rufa* but are used sparingly because of high costs (Campbell and Evans, 1988). Wire cages around individual seedlings have shown utility in deterring animals (Black *et al.*, 1969) but caging was not a practical consideration until the development of ready made plastic seedling protectors.

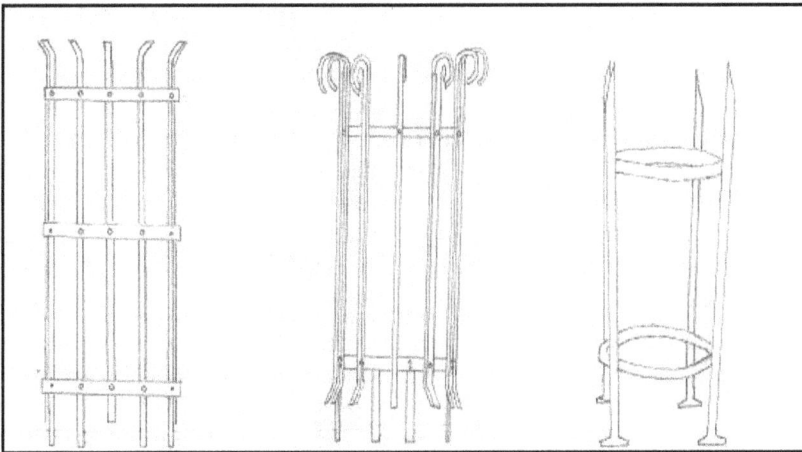

Figure 6.11: Different types of tree trunk guards

A readymade protector of 76 cm long cylinder of photodegradable, polypropylene plastic netting with an inside diameter of 5 cm with mesh opening of 9 mm and strand diameter 1.5 mm is found to reduce gopher damage in pine and fir seedlings by about 77 to 97% and increased the survivability of fir seedlings by 41% (Anthony *et al.*, 1978). Guards made of 20 gauge chicken wire/poultry netting with 1 inch holes are commonly used (Johnson, 1964). Henderson and Craig (1932) indicate that 3 ft high sturdy woven-wire cylinders can be used to protect fruit, shade or other valuable trees from attacks by beaver, *Castor canadensis*.

These cylindrical wire meshes are used effectively against rats (Williams and Hsu, 1979) and Old World porcupines, *Hystrix* spp. (Hartley, 1977) in various trees. Individual wire cages, at least 0.5 m in dia. and 1 to 1.2 m in height, may be used to protect single young trees from deer browsing and antler rubbing (Longhurst *et al.*, 1962). Although found effective these mechanical devices and screens are too expensive and not practical for large scale field projects such as reforestation (Sullivan, 1978). These cylinders may restrict tree development so it is important that the diameter of the guards allows for tree growth and that the guards be removed or enlarged as the trees grow into them (Marsh and Salmon, 1979).

ii. Crown and root protectors

Soil Mounding/Banking: This involves mounding of soil (Wilkinson, 1945), crushed stone or gravel (Ritter, 1978) or heavy manure (Waugh, 1901) on the crown or exposed roots. This technique is rarely used today but it may be worth re-exploring. It is used for protection of ornamentals such as bulbs and occasionally the trees of a backyard orchard. Home gardeners with a severe pocket gopher problem often find this method very helpful (Clark, 1983).

iii. Full tree/ crop protectors

Metals sheets: Metal flashing and other types of shields are used on mature trees to prevent animals from climbing the trunks to defoliate trees or damaging or consuming fruits, nuts or cones. Bands of galvanized metal or aluminum flashing have been used to prevent ground squirrels, tree squirrels (Shubert and Adams, 1971; Powell and Powell, 1977), rats (Williams and Hsu, 1979), raccoons (*Procyon lotor*), woodchucks (Logsdon, 1981) and porcupines (Schemnitz, 1994) from climbing trees and causing damage. For squirrel exclusion, a 2 to 3 ft wide band beginning 2 ft above ground is usually used but Shubert and Adams (1971) and Powell and Powell (1977) reported that 18 inch wide bands as sufficient. The limitation here is one tree which is not protected may allow the animals to climb and provide access for tree to tree travel. Protective wraps such as canvas or 1 inch mesh poultry wire have proved effective in preventing rabbit damage in haystacks (Welch, 1967).

Domes or Caps: Warder (1867) suggested encircling tree stems with inverted funnels made of brown wrapping paper protect them against rabbits. Others have suggested similar domes for protecting seed spots from ground and tree squirrels as well as chipmunks, *Eutamias* spp. (Stoeckeler and Scholz, 1956). Domes are also used to protect forestry seed spot plantings from depredation by small rodents, particularly white-footed mice, *Peromyscus leucopus* (Garlough and Spencer, 1944).

Plant materials: Prickly or physically restrictive plant materials are also used to protect seeds and seedlings from depredating mammals. Barry (1860) suggested securing thorns, briers or some prickly brush around the base of trees to protect the trunk bark from damaging.

iv. Rat proofing

Preventative methods such as rodent-proofing are also humane, as well as an essential and probably under-used component of effective control (Mason and Littin, 2003). Rat proofing is practiced by keeping food grains in rat proof containers, a high plinth of the store room, metal sheeting in lower part of wooden door and netting of windows and ventilators.

6.5. Traps and Snares

The use of snares predates recorded history and used by ancient man indicated in Paleolithic artwork in the range of 25,000 years ago (Boddicker, 1982). According to Bateman (1971) Old Testament chronicles make mention of the Israelites using nets, traps, snares and pitfalls. Phillips (1961) and Bateman (1971) make reference to American Indians using snares to produce furs for the earliest trade with Europeans in the New World. Various kinds of traps are used for the capture of different kinds of animals. The size and type of trap will determine the type of animal / bird to be caught (Morris, 1968; Neal and Cock, 1969; Wiener and Smith, 1972; Maly and Cranford, 1985; West, 1985; Ohgushi, 1986; Schwan, 1986; KeshavaBhat and Sujatha, 1987; Thompson and Macauley, 1987; Slade *et al.*, 1993). This is not only dependent on size-related criteria, but also because different species often display a preference for a particular trap type (Hansson and Hoffmeyer, 1973; Rose *et al.*, 1977; Willan, 1986). Trapping is more art than science, thus people using traps have their own preferences or opinions and modify and develop traps and thus a variety of traps are being evolved. A detailed description of various trap designs is given by Hygnstrom *et al.* (1994).

I. Traps for Birds

Numerous types of traps and trapping techniques have been tested for trapping blackbirds but have not developed to the point where it can be considered as a method of population control, but it does constitute a means of successfully capturing large numbers of birds (DeGrazio, 1964). The place of trapping the birds also influences the trapping efficiency. Birds such as blackbirds and starlings are difficult to trap except when food is scarce in the fields. Trapping cannot be considered a practical means of reducing blackbird populations around rice and corn fields, but is more effective against both blackbirds and starlings around localized areas such as cattle feedlots (Zajanc, 1962). Live trapping, using walk-in type traps on roofs or other isolated sites can be used to remove pigeons. Extensive trapping is recommended as a control measure of bullfinches (Seubert, 1964).

Pigeon traps: Mechanical traps are used to mass trap pigeons in places where they are plenty. The trap is normally made of lightweight aluminum rods spaced one inch apart with two chambers, a small entrance chamber and a larger holding chamber. Water or food materials are kept in the trap and pigeons are allowed to get acclimatize for 2 or 3 days to alleviate trap shyness and then mass trapped (Martin and Martin, 1982). Trapping is a relatively effective pigeon control method which normally can be used. Unfortunately, it can be quite high in terms of labour requirements, but this is usually offset by its safety, selectivity and its low profile (Gilman, 1978).

Australian crow trap: The Miller-type cannon net trap and the modified Australian crow trap have been the two most successful traps used. Australian crow traps can be made in nearly any size but should be at least 8 to 10 ft square and 5 to 6 ft high. Decoys are usually kept and crows get inside by itself seeing the food and decoy. These traps are made attractive for corvids and give the possibility of multiple captures continuously (Tsachalidis *et al.*, 2006). The modified crow trap is more productive for capturing large numbers of birds steadily over an extended period even up to 500 birds in 15 traps in one day. When operated properly, one crow trap is capable of catching 50 to 75 birds per day (DeGrazio, 1964). Modified Australian crow traps are used effectively to protect pecan orchards and watermelon fields from hooded crows, *Corvus corone sardonius* (Moran, 1991).

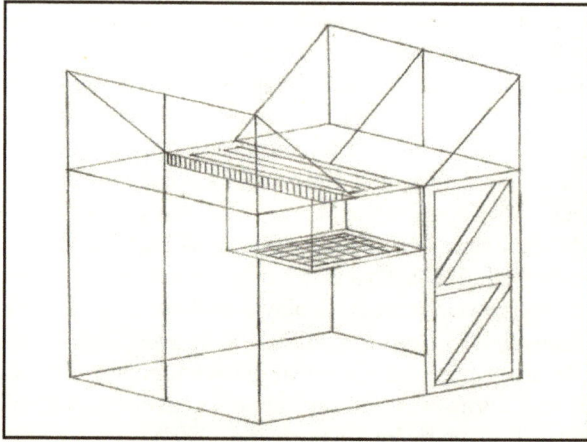

Figure 6.12: Australian crow trap

Starling trap: It is a modification of the Australian crow trap used for capturing crows and magpies. This self-operating cage trap has been found effective for capturing starlings. The trap is simple in principle; the starlings enter through small holes in the welded-wire center section of the V-shaped cage top. Once inside, they endeavor to escape by going to the outer walls rather than through the top openings. This trap can also be used for trapping blackbirds, grackles and cowbirds. For best results, use bait materials on which the birds in the vicinity of the trap are accustomed to eat (Zajanc, 1962). By increasing the trap area volume and placing two or more small traps together also got better results in capturing starlings (Bogatich, 1967).

Parotrap: This trap is used to mass trap birds especially the rose-ringed parakeet, *P. krameri* in the field. The trap measured 3.1x3.1x1.8 m. It consisted of 4 side panels constructed of aluminum frame and plastic-coated poultry wire. The top panels were constructed of wood and galvanized poultry wire. The center plywood roof panel had two parallel slots for entry, 75 cm long by 5.4 cm wide. On the underside, the outer edge of each slot was trimmed with metal flashing to prevent escape. The bait tray constructed of wood and metal screen was suspended 60 cm beneath. Trapping efficiency of parotrap was reported good with an average trapping of 7.16 parakeets/ trap/day near a wheat field (Bashir, 1979). With the installation of four parotraps in eight acre sunflower field the bird damage to sunflower has reduced from 13.4% to 4.8% with an average reduction of 71.6% in nine days (Bashir, 1979). Large numbers of parakeets can be collected by use of the parotraps and consumed locally as food or marketed as pets in Pakistan (Bashir, 1978). But this trap was not found effective in trapping monk parakeet, *Myiopsitta monachus* (Tillman *et al.*, 2004).

Drop-in funnel trap: The main body of the trap measured 4.7x3.1x1.8 m and has a large funnel-style side entry and stepwise baiting system. The funnel made of galvanized poultry wire and wood panels direct the birds from the lower food tray into the main body of the trap. This is used to trap monk parakeet, *M. monachus* in the field but not found very effective (Tillman *et al.*, 2004).

Net trap: Cannon netting is a method of catching a large number of birds. These are nets pulled rapidly by explosively-driven projectiles to cover an area and presumptively capture birds before they have time to escape. They are excellent for capturing pigeons, ducks, geese and turkeys. The cannon net trap has been more productive for capturing large numbers of redwings in single catches. It was reported that in one instance, 1,550 birds were taken by firing two nets simultaneously.

Light trap: A light trap consists of a series of arches of aluminum pipe covered by cotton or nylon netting. The arches taper in height from 35 ft in front to 10 ft at the back and are arranged to form a funnel when covered with netting. A canvas covered, air tight holding cage, approximately 8x8x10 ft, serves as a gas chamber at the small end of the funnel. A battery of five or six, 1000 watt incandescent lights is placed in the holding cage as the attractant. Drives should be carried out on a dark, moonless night to be most effective. Several 'drivers' scare the birds from the branches of the roost trees; at the same time, the lights are turned on to attract the birds into the net. It was reported the light trap catches goes up to 120,000 birds with starlings outnumbering blackbirds (Zajanc, 1962).

Elevator Trap: The elevator trap is a small, portable cage approximately 24" long 16" wide and 8" high with a weighted elevator at one end. A small wire cage with two sides open is affixed to the elevator, with a bait box just beyond to attract the birds. In attempting to get to the bait, the bird must step into the wire cage; its weight forces the elevator down to the base of the trap where the only escape route is into the main part of the cage. As the bird leaves, the elevator returns automatically to its original position. Leaving decoy birds in the trap will attract others. This trap is useful for trapping young starlings during the summer months in orchards and vineyards (Zajanc, 1962).

Sticky traps: Sticky traps are used to trap small birds and rodents from time immemorial. It has been used since the time of the ancient Romans and was first mentioned in the literature as early as 1678 (Bateman, 1971). Liming was the predominant method of trapping birds in Europe from antiquity to the late 19th century when small birds were trapped on a large scale for food and sport (Hogarth, 1929). Sticky substances from natural gums of plants were painted on twigs in areas where the birds were apt to light.

The adhesive materials can easily gum up flight feathers and the bird's light weight makes it more difficult for it to escape entanglements. Plant resins and gums, sometimes in combination with vegetable oils, have a long history of use. Holly bark (*Ilex aquifolium*) was used extensively in old European formulas, slippery elm (*Ulmus rubra*) in America, and various members of the mistletoe family (Loranthaceae), particularly *Loranthus europaeus* in Europe and Africa. Over the world, especially in tropical areas, bird limes have been made from the sap of members of the mulberry family (Moraceae), figs (*Ficus* spp.) and breadfruit (*Artocarpus* spp.) (DeWit, 1967).

II. Traps for Mammals

Various kinds of traps were used for the capture of different mammal pests with limited to high success. Traps for mammals can be of killing type or restraining type. There are five main categories of killing traps in use: deadfall traps, spring traps, snares, drowning traps and pitfall traps (Proulx, 1999; Powell and Proulx, 2003). Deadfall traps use gravity to kill an animal by crushing its skull, vertebral column or other vital organs. There are two types of spring traps; one has spring-powered bars that kill an animal by crushing a vital region of the body, generally the neck; the other has rotating jaws which have two hinged metal frames that allow a torsion spring to rotate the frames in a scissor-like action (Garrett, 1999; Powell and Proulx, 2003). Drowning traps restrain an animal underwater and kill by hypoxia-induced death. Finally, less commonly used traps include pitfall traps with water at the bottom to drown small rodents (Proulx, 1999).

Five kinds of restraining traps are widely used: stopped neck snares, leg-hold snares, leg-hold traps, box or cage traps and pitfall traps (Proulx, 1999; Powell and Proulx, 2003). Neck snares are made of a wire loop set vertically, so that the head of the animal enters the wire loop, which then tightens around the neck of the animal. Leg-hold snares are also made of a wire loop, but placed horizontally and designed to close upon the animal's leg(s) to restrain it. The trap is attached to the ground or anchored by a chain or cable. The anchor restrains the animal by snagging on surrounding vegetation (Iossa *et al.*, 2007).

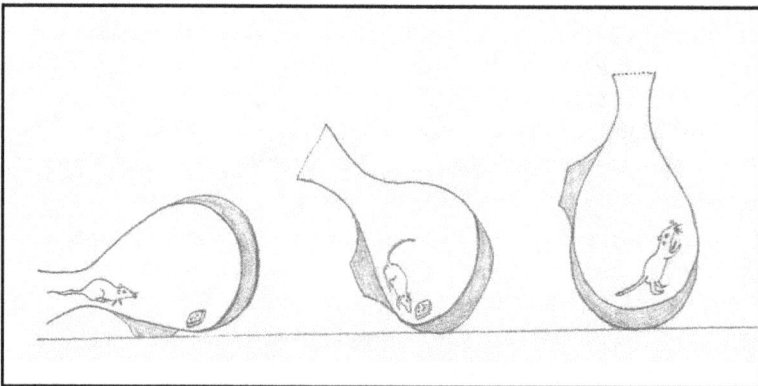

Figure 6.13: Pot trap for rodents

Simple box Trap

Box traps are constructed from a wide variety of materials including plastics, wire mesh and wood (Meyer, 1991; Proulx, 1999). The principle of box trap is when an animal enters the trap through an opening attracted by bait, it triggers a device (eg treadle) that causes the door to close and lock. Box traps vary in size, and their design depends primarily on the target species (Powell and Proulx, 2003). Box traps are humane (Iossa *et al.*, 2007) and widely used for trapping mammals. Further, box traps can capture a range of species, but unlike other trap methods, non-target species are typically released unharmed. The pitfall trap is a smooth sided container buried in the earth. Whilst with killing traps all or the majority of non-target individuals captured are killed, restraining traps vary in mortality rates from 0% in box traps to 17% in leg-hold snares (Logan *et al.*, 1999; Potocnik *et al.*, 2002). Because of many advantages of restrain traps over death traps, these traps are described in detail here under.

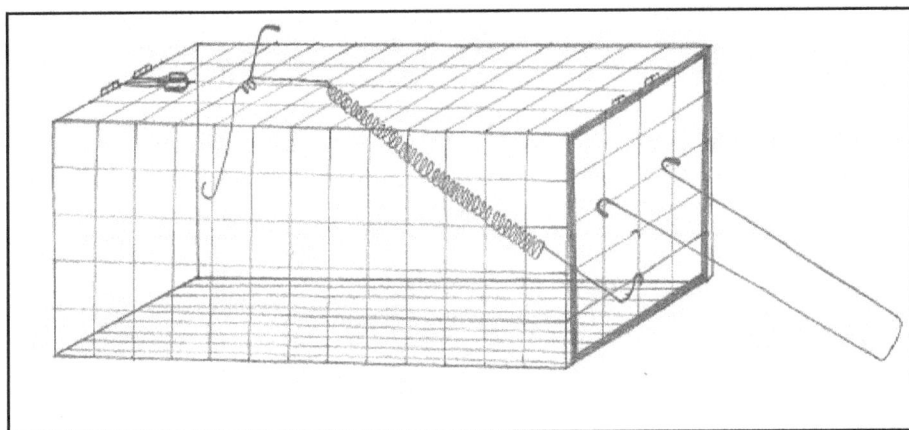

Figure 6.14: Box trap for rodents

Cage/ Pen Trap

Cage or pen traps are based on a holding container with some type of a gate or door (Mapston, 1999). Trapping with the box or similar type trap is often effective, but this is a slow method of control for cottontails. A suitable cage trap and baiting system has been developed for rabbits. Trap spacing and inter-bait distances are based on behavioural studies of rabbits grazing cereal fields (Cowan *et al.*, 1989). It is not a successful method for controlling jack rabbits because of their reluctance to enter a trap or dark enclosure (Johnson, 1964) but highly successful for rodents. In general, cage traps, including both large corral traps and portable drop-gate traps are most popular and effective, but success varies seasonally with the availability of natural food sources (Barrett and Birmingham, 1994). Trapping in conjunction with rabbit drives, where a large territory is surrounded by men and the rabbits driven into a corral, may be considered a method of trapping that has been successful. Another similar type of trapping operation is the construction of a small corral along a fence surrounding a protected field, funneling the rabbits to a one-way gate into the corral (Johnson, 1964). Some animals like pigs are large and powerful, and trap materials and construction techniques must be able to withstand the forces exerted by captive animals. At a minimum, side panels should be constructed of 4 gauge welded fencing or its equivalent and even heavier materials should be used for gates and frames (Barrett and Birmingham, 1994). Delayed triggers should be used so multiple animals can enter before the door trigger is engaged, thus increasing the probability of capturing multiple animals each time (West *et al.*, 2009).

Longworth trap: The Longworth trap is the most commonly used small mammal live trap. The design has been around a long time (Chitty and Kempson, 1949) and a wealth of hints and tips now exist (Gurnell and Flowerdew, 1990). Longworth traps come in two parts (nest box and tunnel), which fit inside each other for ease of transport. Each trap is of 14x6.5x8.5 cm when broken down (the tunnel fits inside the nest box). The small diameter of the tunnel (5.0x6.2cm) limits the catch to animals less than 700g. The box serves as a refuge for the captured animal.

Larger animals are harder to catch with a standard longworth trap, because larger animals might get their back, hind feet, or tail caught under the door when the traps spring and thus can back out to avoid being trapped (Boonstra and Rodd, 1982). The Living Trip-Trap also looks like longworth trap but made of plastic and unlike longworth, the captured animal can be seen without opening the trap (Barnett and Dutton, 1995). Cockrum (1947) reported that live traps (8"x3"x3") caught 2 to 3 times more small mammals than snap traps Live traps were reported to catch more shrews (*Sorex araneus*) and mouse (*Apodemus flavicollis, A. sylvaticus*) but not voles (*Microtus agrestis* or *Cleithrionomys glareolus*) (Hansson and Hoffmeyer, 1973).

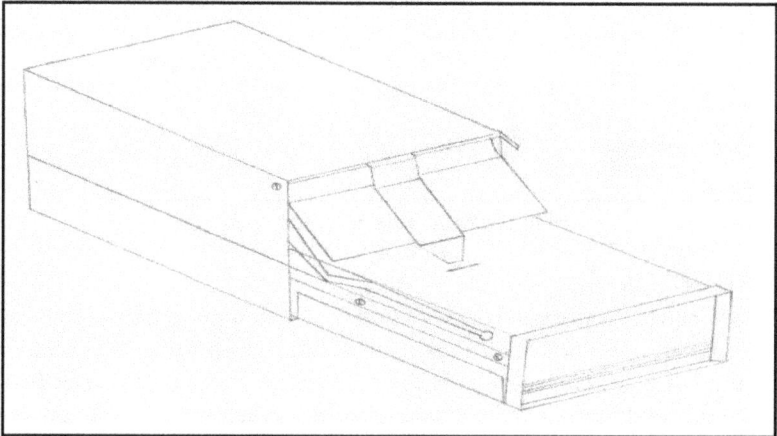

Figure 6.15: Longworth trap

Sherman, Havahart and Snap trap: Shermans are simple box-shaped collapsible traps designed for the capture of animals with long tails. These are made of metal sheets whereas Havaharts are of strong wire mesh. The disadvantage is that their open construction offers the captured animal no protection from the weather, a potential source of increased mortality of captured animals (Barnett and Dutton, 1995). Sealander and James (1958) reported that Sherman live traps as more effective in catching small mammals than the Museum Special trap, a snap trap but vice-versa by Wiener and Smith (1972). Alexander *et al.* (1987) compared the efficiency of Sherman live traps and Japanese plastic snap traps in trapping 2 sympatric shrew species and found that Sherman traps caught significantly more shrews than snap traps. Faust *et al.* (1971) pointed out in particular, that Sherman live traps were inferior to Museum Specials in catching shrew, *Blarina brevicauda*. Snap trap recorded higher capture rates of pacific rats, *Rattus exulans* while a reverse-bait trigger rat trap captured more black rats, *R. rattus* and Norway rats, *R. norvegicus*. The reverse-bait trigger rat trap was found to be efficient in killing larger rats (>100 g) whereas the snap traps are good for smaller rodents (Theuerkauf *et al.*, 2011).

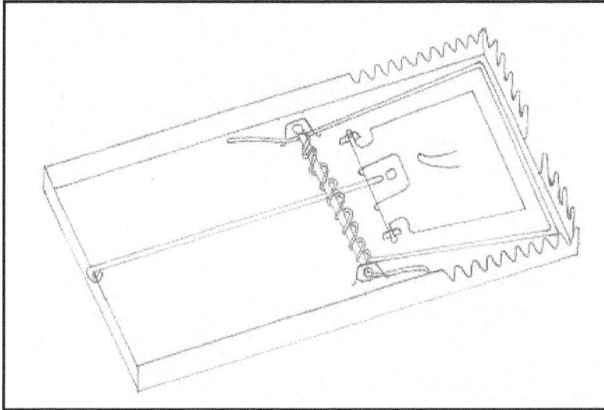

Figure 6.16: Snap trap for rodents

Arboreal traps/ aerial traps: The capture success of harvest mice, *Micromys minutes* and dormouse, *Muscardinus avellanarius* were increased substantially by fixing traps approx. 1 and 3.5 m above ground levels (Morris and Whitbread, 1986). The live capture of the bamboo rat, *Kannabateomys amblyonyx* was only possible by placing traps off ground on bamboo bridges (Kierulff *et al.*, 1991). It was reported that 20% of woodmice, *Apodemus sylvaticus* and 17% of dormouse, *M. avellanarius* were captured off the ground (Tattersall and Whitbread, 1994). Tree squirrels are often trapped by arboreal traps fixed in tree branches.

Broma traps: Broma traps are used to capture bush pigs, *Potamochoerus porcus*. Crop plants are fenced by pig-proof wire providing an entrance through a gate into a trap, which could be closed off once they had entered to feed and found successful in trapping the pigs. The pigs were found reluctant to enter through a one meter gate and when the gate was extended it was found very successful. A portable model of broma traps are made available which can be placed at places of heavy damage (Grange, 1986).

Foothold traps: Trapping especially using leg hold traps is reported to be the effective method to capture badgers (Todd, 1987). Foothold traps and body grasping traps are commonly used for beaver, muskrat and smaller mammals like ground squirrels, rats and marmots.

Foot hold traps are effective in trapping Culpeo fox, *Pseudalopex culpaeus* and feral cats, *Felis catus* and aids in their management (Travaini *et al.*, 2001; Wood *et al.*, 2002). Like other traps there is a potential of capturing non-target species which can be lowered by using proper trap size, pan tension devices, break-away mechanisms, species specific baits and selecting trap locations (Conover, 2002).

Glue traps: The use of glues against mammals is not very effective nor in widespread use. It is interesting to note that the tribal Indians reportedly spread the juice of the peepul tree, *Ficus religiosa* on leaves along jungle trails. Wild animals including tigers, attempting to get the sticky leaves off their paws would become blinded as they rubbed their faces with their paws (Burton, 1918). The most widespread use of sticky materials is the use of 'glueboards' to catch rats and particularly mice.

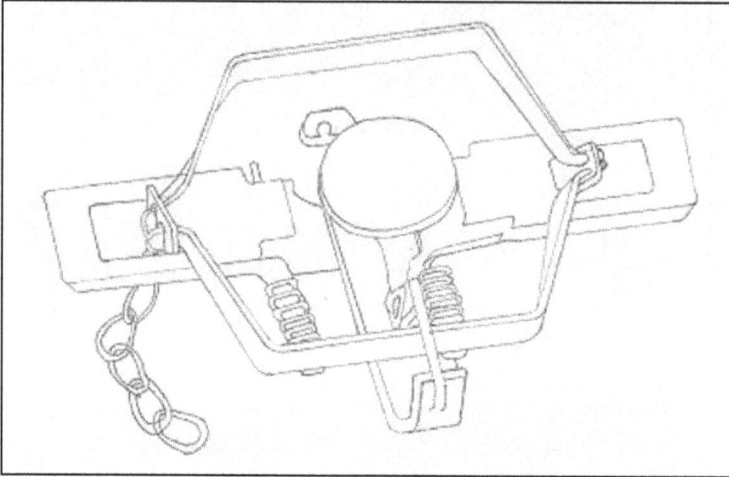

Figure 6.17: Foothold Trap

Sticky-boards coated with a water-resistant adhesive and a rodenticide has been used successfully to trap sewer rats but cannot be used in large scale (Brooks, 1962). Rat glues were used as early as the 18th century (Holland, 1802). Early materials were made from latex and gums of many trees but current ones also use industrial chemicals like polyethylenes and polybutenes. Hockeynos (1958) in his excellent studies on the chemistry of entanglements reported the best rat glue formula as, Pine resin (31%), Abalyn* (18%), polyethylenes (5%), latex (6%) and mineral oil (40%). Most of the current rat glues are modifications of Hockeynos pioneer work. Glueboards have much the same advantages of mechanical traps. There are no toxic chemicals to endanger pets or children. They have most of the advantages of traps but have some disadvantages. Their use may be limited by temperature, moisture, dust, vapours, etc. Dampness creates a film on the surface of the glue so rodents can skate over it. Dust, sand and grease also coat the glue surfaces, lowering their effectiveness. They are probably more expensive than mechanical traps as they must be discarded when they have caught animals or been exposed for long periods of time. In some situations, rats are able to pull out if only one or two feet are caught (Fitzwater, 1982).

Pit fall traps and trenches: The most common measures to avoid damage by bush pig, *P. porcus* are making pit fall traps and trenches (Kagoro-Rugunda, 2004). Farmers around Kasohya-Kitomi forest reserve in Uganda, dig trenches to a depth of one meter so that large animals do not cross to the farms from the forest (Fungo, 2011). Pit-traps have also been known to be used against crop raiding elephants in their paths approaching to the field (Hoare, 2001a). But elephants have also been known to fill them in by kicking soil from the edges into the trench, thereby filling it and enabling them to cross (Sukumar, 1989).

Snares: Snares can also be effective but should be used cautiously in areas where livestock, deer or other non-target animals are present. The Australian brush-tailed opossums were reported to be controlled by using snares in New Zealand (Howard, 1964). Feral pig, *Sus scrofa* were effectively controlled by snaring in Hawaiian Islands. In high pig density areas (14.3 animals/km²), 7 worker hours/pig was utilized to remove

175 animals whereas it took 43 worker hours/pig to remove 53 pigs in less dense area (6 animal/ km²).

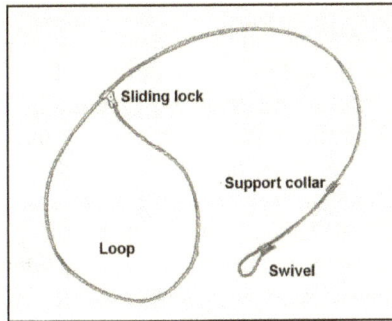

Figure 6.18: String snare

Special trappings: Some reports say that boars could detect oestrous sows and were attracted and actively seek them out (McGlone and Morrow, 1987; Rathore, 1989). A study was conducted to compare the conventional trap and trap with oestrous sows in trapping wild pigs. The oestrous sow was kept in weld-mesh panel traps that in turn surrounded by a round silo trap equipped with two swing-gate entrances. But oestrous sows did not found to increase the trapping efficiency (Choquenot *et al.*, 1993).

The 'Judas' concept is the use of radio-telemetry to find and control flocks of feral goats (Taylor and Katahira, 1988) and pigs (Wilcox *et al.*, 2004). In essence, the technique relies on tracking radio tagged animals to locate larger social groups, which can then be removed using a variety of methods. This technique has been used with some success in Australia (McIlroy and Gifford, 1997) and in United States (Wilcox *et al.*, 2004; McCann and Garcelon, 2008). Once captured, individuals are equipped with radio transmitting equipment (Braun, 2005), visibly marked with paint or ear tags and released. Subsequently, the tagged individuals will reintegrate with social groups and enable managers to locate large congregations.

Gopher traps: Many special gopher traps have been devised but two kinds, a choker type and a double-pronged snap type are widely used. A box-type choker trap is mainly used in many citrus orchards where gophers are a problem of greatest economic importance. Traps should be set in an open, main-traveled burrow, not in the lateral tunnels which lead to surface mounds or feeding holes. It is best to set two traps, one facing each direction in the tunnel. Traps should be firmly placed, well into the tunnel with as little disturbance as possible. The opening made in order to set the trap should then be covered with sod or clods so that little light reaches the trap to achieve successful trapping (Cummings, 1962).

Though trapping is a time consuming and laborious process, it has many advantages than any other management option. Unlike poisoning, trapping does not have any risk of secondary poisoning (Bowie and Ross, 2006) however traps also put non-target fauna at risk (by catching) even when live traps are used (Waldien *et al.*, 2004) but the non targets can easily be released back. Trapping and relocation is also a common practice which is accepted by all and offers conservational management. However relocation is always not feasible. Trapping of black bears using culvert trap and relocated to distant places was not successful because of the homing behaviour of the bears and they start homing as soon as they are released (Barden *et al.*, 1993).

Placing a trap is also a unique technique which determines the success at a large. Small mammals prefer to run along the edge of things (edge effect), and will generally run round the edge rather than across it, so traps should be placed accordingly. The success of trapping heavily depends on the selection of trap and procedure of installation but also on the age, activity and behaviour of the animal. For example the raft traps were found very effective than the land traps for capturing coypu, *Myocastor coypus* (Baker and Clarke, 1988). Age of pig significantly effect on trap success and a decline in trapping occurred in areas of older pig activity (Fox and Pelton, 1977). But Daly (1980) reported that adult rabbits were most trappable than young ones and females than males.

Trapping is the most common and successful method for the management of Cape otter (*Aonyx capensis*), honey badger (*Mellivora capensis*), musk cat (*Genetta* sp.), wild cat (*Felis libyca*), black-footed cat (*Felis nigripes*) and all rodents (Hey, 1964) because of their behaviour. An intense trapping program can reduce populations by 80 to 90%, but that some individuals are resistant to trapping. Thus, trapping alone is unlikely to be successful in entirely eradicating populations (Choquenot *et al.*, 1993). It is very difficult to assess the number of traps to be used until trapping success is known. Gurnell and Flowerdew (1990) consider that, as a rule of thumb, if 50 to 60% of traps are filled at one time then more traps should be put down. Success of trapping depends on the efficiency of the persons handling them. In the hands of inexperienced hunters traps kill numerous useful and harmless creatures.

Chapter 7

Electrical Scaring

An electric shock can be used to scare any living organism but should be used properly to avoid risk of killing non-targets. In situations, electrical fencing may be considered as beneficial and economical than ordinary fencing. Electrical perching scares bird pest in the field and electrocution traps are used to capture and kill rodents and other pest animals.

7.1. Electric Perch, Electric Collar and Electrocution Traps

Electrical perching

Electric shock systems are available to repel birds from ledges or similar sites. Birds that land on an electric circuit closes the electrical connection resulting in a repelling but harmless electrical shock (Johnson, 2000). An electric perch can be made by suspending two wires, spaced 2 to 2.5 inches apart, some 10 to 14 ft above a grain plot. Such wires be suspended over the entire plot at 25 yard intervals and thus at least a 15,000-volt transformer to be used (Pfeifer, 1956). If more than 2,000 ft perch line is to be built, then a 60 milliampere transformer is recommended (Pfeifer, 1957).

This electric perch was tested over a small (49x69 ft) isolated area of sunflower plots. Two wires, separated about 2.5 inches by porcelain spacers and charged with 15,000 volts were suspended about 14 ft above ground level, across the middle of the area. Use of this apparatus provided only partial protection from sparrows and finches and was completely ineffective against red-winged blackbirds. Furthermore, some of the installations of this device have been so destructive to beneficial birds, particularly doves, that the users have been forced to cease their operation (Chubb, 1959). But, Zajanc (1962) reported electric perches as effective against blackbirds at a distance of 50 yards and provided control of bird damage at a minimum cost.

Electric collars

Shock collar or electric/electronic collar is that which deliver electrical shocks of varying intensity and duration to an animal via a radio controlled electronic device incorporated into a device fitted on its neck. Electric shocks have been used to establish aversions to visual and gustatory stimuli in albino rats (Krane and Wagner, 1975). Electric collars and electronic ear tags are used to train domestic cattle to avoid certain portions of pasture (Quigley *et al.*, 1990; Tiedemann *et al.*, 1999). Shock collars appear to be effective in modifying foraging behaviours and/or movements in captive coyotes (Linhart *et al.*, 1976; Andelt *et al.*, 1999), captive wolves (Shivik *et al.*, 2003), island foxes (*Urocyon littoralis*) (Cooper *et al.*, 2005) and wild wolves (Schultz *et al.*, 2005; Hawley *et al.*, 2009; Rossler *et al.*, 2012). Linhart *et al.* (1976) successfully used electronic collar to condition coyotes to avoid black domestic rabbits. An electric collar was tried to create aversion in lamb depredating coyotes (*C. latrans*) for 22 week period and reported to avert all the attempts of attacking a lamb and caused coyotes to avoid and retreat from the lambs for >4 months (Andelt *et al.*, 1999). Shock collars were found to condition wolves, *Canis lupus* for forty days post treatment to avoid baits (Rossler *et al.*, 2012). In a study, wolves were fitted with radio collars and domestic calves with a battery operated system, which gives an unambiguous cue of shock if a wolf approaches within 1 m of the calf, making the predator to withdraw from the prey (Shivik and Martin, 2000). Use of electric shock as aversive stimuli to alter animal behaviour has been studied in field situations also (Krane and Wagner, 1975; Linhart *et al.*, 1976; Quigley *et al.*, 1990; Tiedemann *et al.*, 1999). Shock collars restricted wolf access to specific sites (Hawley *et al.*, 2009). Schultz *et al.* (2005) found shock collars prevented free-ranging wolves from visiting a study farm. Thus shock collars could serve as a useful non-lethal tool for managing livestock depredations, particularly in chronic problem areas and with endangered populations (Rossler *et al.*, 2012). Shock collars may be especially useful and perhaps only practical with 'high value' animals (e.g., grey wolves, grizzly bears). Implementation, use and maintenance are expensive and a disadvantage is that shock must be delivered precisely (Mason, 1998).

Electrocution traps

These devices consist of an open ended box baited with dry food. The floor is made of two plates which are terminals; the animal especially rodents bridging these two plates receives a 2 min-long shock, transmitted via the feet, of around 2000V. These traps are usually of battery-driven, hence easy to power and also portable. It is comparable with poisoning and work faster than snap traps. Their disadvantage is that, they have to be reset between kills. Mice may also sometimes move too fast to make a good contact between the plates and also found that three out of five rats fell over when shocked, broke the contact and so failed to be killed. They are found effective against rodents and possums but not economical (Mason and Littin, 2003) and also unacceptable on humanitarian grounds (Dix *et al.*, 1994). However, electrocution traps are reported to kill rodents swiftly and with little distress (Mason and Littin, 2003). The time taken by the animal to die is very brief (~2 min) and the escaping animal being shocked are not left with burns and thus can be considered for management of animal pests especially rodents.

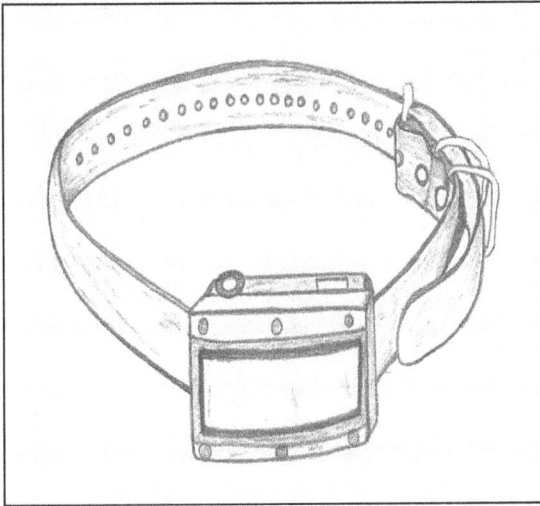

Figure 7.1: Electric collar belt

7.2. Electrical Fence

Electric fencing for birds

Electrified wires providing nonlethal shocks have been used as a repelling tactile stimulus to deter pest birds. Although operating on high voltages, they are not lethal because of low amperages. Electrified wires have been used to prevent damage by birds to agricultural crops (Pfeifer, 1956; 1957; Zajanc, 1962). The birds must come into direct contact with the charged wires in order to be repelled and this proves to be the major limiting factor in their usefulness (Hygnstrom and Craven, 1988). Electric non-lethal shocking devices for repelling birds have generally received little attention. The approach probably deserves greater consideration for special situations.

Electric fencing for mammals

Electric fences are cheaper than conventional fences and because of this; they are being used increasingly to manage mammals (McKillop and Silby, 1988). In these fences, electric current is supplied by low-impedance, high voltage chargers, which provide timed pulses (45-65 /min) of short duration (0.0003 sec.), followed by a relatively long period without current flow. The short-duration, high-energy pulses provide sufficient energy (> 3,000 volts) to deter deer, while still allowing an adequate period without current to allow animals to free themselves from the electrified wires. Plug-in and battery or solar-operated chargers are available to maintain in excess of 5,000 volts on several kilometers of fencing. Simple electric fence designs can be used to deter deer damage where deer numbers are low to moderate and other food sources are available to deer. An example of a simple electric fence design is the single strand electric fence. Several (six- or seven-wire) high-tensile fences have been found to effectively control deer damage for small areas (McAninch *et al.*, 1983). Electric fencing with electrified high-tensile steel wire and electrified polytape or polyrope is used to manage white-tailed deer damage (VerCauteren *et al.*, 2006). For larger areas, an electric fence is an unfamiliar object for deer exclusion and a deer investigating it for the first time often will touch the fence with its nose (Prior, 1983). However, a deer foraging at night may not see the fence and could touch the wires with its neck, back or chest (Tierson, 1969). If an animal has almost crossed the fence before an electric pulse is generated, it will likely complete the crossing. Deer are reported to have learned to avoid receiving shocks by jumping through electrified fences (Tierson, 1969).

Simple electric fences with only three strands and a voltage of 5.5 kV has been successful in controlling damage causing animals over very large areas in Kenya, but requires a very active community involvement and a full time fence attendant (Omondi *et al.*, 2004). Electrical fencing is reported as the efficient and cheaper method for reducing crop damage by nilgai, *Boselaphus tragocamelus* (Goyal and Rajpurohit, 2000) and blackbuck, *Antilope cervicapra* (Chauhan and Singh, 1990). Power fencing along certain strategic areas or around highly vulnerable crop fields can check the stray wild pigs (Chauhan, 2011). The only recommended way to reduce raccoon damage is electric fencing. Muskrats and porcupines can be repelled by a combination of net wire and electric fence (Fitzwater, 1972). Electric fences are also used to reduce rat damage in rice crop (Shumake *et al.*, 1979). Sturdy wire mesh fencing, particularly with the addition of an electrified wire about 6 to 8 inches off the ground, seems to be the most effective fence design to exclude wild pigs (Hone and Atkinson, 1983). Reidy *et al.* (2008) found that simple electric fences consisting of only 2 electrified wires, one at 8 inches and one at 18 inches above the ground were as effective as 3-wire designs and reduced daily intrusions of pigs into an area by 50%. Electrical, smooth-wire fence designs are not complete physical barriers but rely on electric shock to make the animals to avoid the fence (McKillop and Silby, 1988). In order to protect apiaries from black bears, *Ursus americanus* electrical fences of relatively low secondary voltages were used and not found much effective (Stoner *et al.*, 1938). Electic fences were reported successful in deterring Asiatic bears, *Ursus thibetanus* from entering apiaries (Huygens and Hayashi, 1999). Floyd (1960) indicated that electric fences were not consistently effective if bears had entered apiaries and discovered honey. High-voltage electric fences are generally effective in excluding black bears from apiaries if vegetation is not permitted to grow around the wires and a 24 inch 'chicken netting' apron is used at the bottom of the fences (Johansen, 1975). A 30 cm apron kept below the ground is necessary to exclude rodents and rabbits and in areas favoured by digging of rabbits, aprons up to one meter is useful (McKillop *et al.*, 1998).

Electric fencing is widely seen as the only nonlethal option for reducing crop damage by larger animals like elephants (Taylor, 1999). Electric fences are found to be highly effective at preventing crop damage by elephants, *Elephas maximus* in Assam (Davies *et al.*, 2011). Electric fencing is increasingly used as a tool for African bush elephants, *Loxodonta africana* conservation in human-dominated landscapes and there are few empirical studies to demonstrate that electrified barriers are effective in deterring elephants from raiding crops. In Kenya, electrical fencing is proved to be very effective to solve elephant problems (Nelson *et al.*, 2007; Omondi *et al.*, 2004). But many have stated that elephants are capable of going through the most sophisticated barriers also, including highly electrified fences when they are determined to do so. Kioko *et al.* (2008) examined the effect of intact fence wires, presence of current and amount of voltage on fence breakage by elephants. The mere presence of current did not minimize fence breakage by elephants whereas intact wires do. Different types of electrical fencing for elephant exclusion (Grant *et al.*, 2007) are given in table 7.1

Comparison of different electric fences showed no clear relationship between effectiveness of fences and factors such as design, construction and voltage. Some high-specification fences have proved ineffective against elephants, while other simple fences

Table 7.1: Different types of electrical fencing for elephant excusion

Fence type	Height (m)	No. of electrical strands	Total no. and type of other strands
Legal for elephant containment	2.4	3	20
Recommended by owners & managers	2.4	4	17
Electrified Armstrong fence	2.4	6	6 cables
21 strand fence	2.4	5	21 steel wires
19 strand fence	2.4	4 at shoulder ht	19
17 strand fence with wire netting	2.4	5	17 (mesh for 1.2 m)
6 strand fence	2.0	3	6
3 strand fence	2.0	3	3

have worked well for a long period. The previous experience of elephants with electric fences in a particular area and shooting of fence-breakers are considered to be more important than any design criteria. Electric fences should be thought of as demarcations of 'no-go' areas for elephants rather than as real physical barriers (Thouless and Sakwa, 1995).

The disadvantage of this fencing is some predators may gain access through fences by jumping, crawling through or digging under. If they are contained within fences, they may resort to excessive damaging. As a consequence of these factors, exclusion fences are not consistently effective in preventing or reducing predation (Wade, 1982). Sometime birds collisions over these fences occur and not only kill the birds but also ground the wire and make short circuit. Visual scarers on fences have greatly reduced these collisions and thus may reduce wildlife impacts and increase barrier fence longevity (Baines and Andrew, 2003). For safety reasons, electric fences should always be adequately marked with warning signs and barbed wire should never be electrified (Curtis *et al.*, 1994a).

Chapter 8

Chemical Management

The last sort of vertebrate pest management is by using chemicals but still is the easiest, cheapest and widely used management method. Chemicals are used as repellents, antifertility agents including hormones, tranquilizing/ immobilizing agents and at last the toxicants. Toxicants, even found very effective should be used with caution because of their environmental effects. Some toxicants like amino-pyridine extensively used for bird pest management also act as a bird repellent. When pest bird ingests this chemical, it emits distress cries while flying in erratic patterns before dying, which makes the rest of the damaging flock to fly away from the targeted fields (Timm, 1994a). Effective, species specific and less persistent toxicants can be used for better pest management.

8.1. Chemical Repellents

Earlier people use plant extracts like veratrum as seed repellents against crows, starlings and other birds in corn (Benson, 1966). Many synthetic chemical pesticides (kocide, imidacloprid), food additives (methyl anthranilate, methyl cinnamate), naturally occurring plant defense compounds (pennyroyal oil, caffeine) and others (antraquinone, anthracene, anthrone) were also tested for bird repellence and found some effect (Avery *et al.*, 1994, 1996 and 1997). Compounds such as sodium silicofluoride, lye, creosote and lime-sulfur have been used in runways and burrows of rodents to discourage activity. Nicotine sulfate, oil of citronella, coal tar and a variety of other substances has also been mentioned as objectionable to pest animals but not very effective to repel them. Predator animal scents especially urine or their synthetic analogues applied or kept in scent darts as area repellents have met with little success. A good repellent chemical should give a strong repellent stimulus to deter further sampling of the available food. A repellent which creates a conditional aversion as it is reported in methiocarb for blackbirds by a post ingestional malaise (Conover, 1984) is of much use than instant repellants. In the past, there had been a considerable variety of repellents recommended against mammals in the form of paints, smears or sprays. Many of these afford only temporary protection and must be renewed too often to warrant their use.

I. Synthetic Chemical Repellents

Methiocarb

Methiocarb [3,5-dimethyl-4-(methylthio) phenol methylcarbamate] was originally developed as an insecticide, but soon regarded as a potential bird repellent (Hermann and Kolbe, 1971). Schafer and Brunton (1971) determined that methiocarb as an efficacious bird repellent, reporting low R_{50} values for several bird species. Tobin and DeHaven (1984) reported an R_{50} of 7.3 g/L of water to European starlings (*S. vulgaris*), 2.8 g/L to American robins (*Turdus migratorius*) and 0.3 to 0.9 g/L to house finches (*Carpodacus mexicanus*) and found very effective in the laboratory. Though the LD_{50} of methiocarb is 16.8 mg/kg, a dose as low as 0.1% is reported enough to produce taste aversion in rose-ringed parakeets, *P. krameri*. Birds conditioned to avoid treated grains at 0.1% concentration were subsequently repelled at concentrations as low as 0.01%. It is thus concluded that if the initial application effectively repelled the birds, a lower level of methiocarb might be adequate at subsequent applications (Hussain *et al.*, 1992).

Guarino (1972) suggested that bird damage to corn, soybeans, rice, sorghum, cherries and grapes could be reduced by spraying with methiocarb. Methiocarb was also successfully tested on baits as a frightening agent in feedlots (Schafer, 1978). Rogers (1974) described methiocarb as a conditioning repellent that has post-ingestional effects causing food aversion in birds. Rooke (1984) demonstrated the conditioned aversion of grey-breasted white-eyes, *Zosterops lateralis* in grapes sprayed with methiocarb. The birds did not use a visual cue to distinguish between treated and untreated grapes because no aversion was seen for a visual mimic, calcium phosphate, but probably used a taste cue. White-eyes that had previously eaten grapes formed a specific aversion while those that had not eaten grapes formed a nonspecific aversion suggesting a border treatment for new birds and entire crop treatment for experienced ones. An alternate-row treatment of methiocarb @ 1·7 kg/ha as bird repellent was evaluated for protecting entire cherry orchards from damage by European starlings, American robins, house finches, common grackles and other birds and found greater advantage to unsprayed areas (Tobin *et al.*, 1989). Based on different experiments, Crase and DeHaven (1976) concluded that methiocarb could be an effective broad-spectrum bird repellent and crop protectant.

Bollengier *et al.* (1973), Stone *et al.* (1974) and Jackson *et al.* (1978) reported that bird damage to blueberries was significantly lower in methiocarb treated fields than in the untreated fields. Methiocarb (Mesurol®) treatments to vineyards were reported efficacious against birds (Crase *et al.*, 1976). Bailey and Smith (1979) obtained twice the yield on vines protected from blackbirds, *Turdus merula* and white-eyes, *Z. lateralis* by spraying methiocarb. Mesurol @ 3g/L applied to grapes provided significant protection from depredating robins, starlings and red-winged blackbirds. Even in an early-ripening and highly-vulnerable variety, losses were reported more than halved. In other varieties, percentage losses in untreated units were more than eight times those in corresponding treated units (Kassa and Jackson, 1979).

Methiocarb is also regarded as a repellent to protect corn and rice seeds from birds and rodents (Johnson *et al.*, 1982; Holler *et al.*, 1983). Lefebvre (1978) found in laboratory feeding preference tests that methiocarb repelled fox squirrels, *Sciurus niger*. Failures of methiocarb in protecting cherries (Stickley and Ingram 1973) and blueberries (Dolbeer *et al.*, 1973) from bird depredations are also reported. Methiocarb (Mesurol 75 WP) was not found very effective in reducing fruit losses by frugivorous birds in highbush blueberry, *Vaccinium corymbosum* (Vincent and Lareau, 1993). Methiocarb @ 4.5 kg a.i/ ha did not found to repel birds like house finch (*C. mexicanus*), starling (*S. vulgaris*), robin (*T. migratorius*), scrub jay (*Aphelocoma coerulescenis*), brewer's blackbird (*Euphagus cyanocephalus*) and mourning dove (*Z. macroura*) from damaging ripened fig. It is also suggested that by piercing (starling and kingbird) or tearing away (house finch, robin and blackbird) of the skin, the bird is exposed to little or no repellent action of methiocarb. The bird can then eat the meat of the fig without any ill effect, thus not effective in the field (Crabb, 1979).

Anthraquinone

Anthraquinone apparently acts as a conditioned-aversion repellent with birds. Birds ingesting food treated with anthraquinone become slightly ill and develop a post-ingestion aversion to the treated food source. Birds visually identify anthraquinone in the UV light spectrum and become conditioned to avoid the treated food source. Because of this, anthraquinone use does not require treatment of the entire field, but only areas where birds are grazing. It is found as an effective repellent of blackbirds under captive and field conditions (Cummings *et al.*, 2002; Avery *et al.*, 1998). Anthraquinone at 0.1, 0.5 and 1.0% (g/g) was found to deter consumption of cowbirds on treated millets in caged experiment. Treated birds lost their body mass in all the doses in no choice test and some mortality was noticed in 0.5 and 1.0% treatments. Blackwell *et al.* (2001) used anthraquinone as a corn feeding deterrent to sandhill cranes, *Grus canadensis* in pen trials. Cranes were reportedly consumed 8.6 times more corn in the untreated pens compared with corn treated with anthraquinone. Anthraquinone is regarded as an effective bird repellent for the protection of pine and rice seeds (Royall and Neff, 1961). Dickcissels (*Spiza americana*), a pest of rice and sorghum crops was fed with rice treated with anthraquinone (0.5%) reduced consumption by 70% with respect to pretreatment consumption. In choice tests, 0.05 and 0.1% anthraquinone was found very effective reducing the feeding rate by 90% (Avery *et al.*, 2001).

Flight Control™ (50% anthraquinone) at 4.5 L/ha repelled Canada geese, *Branta canadensis* in terms of bill contacts by 60.6% when compared to untreated control for seven days in captivity. The mean number of geese per observation were also greater on untreated plots (2.6) compared to treated plots (1.4). Residues of anthraquinone were found to dissipate in turf up to 89.1% in one week period (Dolbeer *et al.*, 1998). A combination of Flight Control™ (50% anthraquinone) at 2.3 l/ha and Stronghold™, a plant growth regulator was reported to reduce the mean goose presence in turf by 62.1 % and bill contacts per min by 87.8% and no abatement in repellency was observed even at 22 days post treatment (Blackwell *et al.*, 1999). Anthraquinone and TMTD were found highly effective in minimizing bird depredations to field crops (Welch, 1967).

The effectiveness of a combination of a secondary repellent, anthraquinone (Avipel®) (0.1%), and a primary repellent, d-pulegone (0.17%), in deterring consumption of cereal-based pellets by captive kea, *Nestor notabilis* was studied and a 69% decrease in the food consumption noticed. When birds were then presented with food containing only the primary repellent, there was a further decrease in consumption of 88.7% reduction from initial consumption rates and when untreated food was given it slowly reaches full consumption rate. Thus after the combination product, at least continuous use of the primary repellent is required to maintain the aversion response on subsequent exposure to pellets (Orr-Walker *et al.*, 2012).

Methyl anthranilate

Methyl anthranilate is a well know bird repellent, when applied to seeds or grasses inhibits birds from feeding and repels a variety of birds, such as red-winged blackbirds, starlings, pigeons, jungle fowl, herring gulls, ring-necked pheasants and Canada geese (Avery and Decker, 1994; Avery *et al.*, 1995; Cummings *et al.*, 1995; Glahn *et al.*, 1989; Mason *et al.*, 1989; Marples and Roper, 1997). Both dimethyl and methyl anthranilate were strongly avoided by captive mallards and Canada geese when birds were offered both treated and untreated grain. In the pen trial, 59 kg/ha of the chemical applied reduced goose activity on treated grass plots for less than 4 days (Cummings *et al.*, 1992). Methyl anthranilate was found to deter sandhill cranes, *G. candensis* from eating corn seeds in pen trials. The cranes consumed 9.8 times more untreated corn compared with corn treated with methyl anthranilate (Blackwell *et al.*, 2001). Methyl anthranilate is used for dispersing migrating swallows and killdeers though aerosol applications also (Engeman *et al.*, 2002).

Askham (1992) reported that methyl anthranilate applications reduced bird damage to sweet cherries and blueberries by 43 to 98% and 65 to 99%, respectively, compared with untreated controls. Bird damage was reported slightly reduced in methyl anthranilate treated grapes and sweet cherries but not in blueberries (Curtis *et al.*, 1994b). Avery *et al.* (1992) evaluated two methyl anthranilate formulations in a 0.2 ha aviary enclosure containing blueberries. Neither formulation protected blueberries from foraging by penned cedar waxwings, *Bombycilla cedrorum* despite found aversive in laboratory caged trials. Methyl anthranilate applied at 16 kg/ha did not reduce bird damage in blueberries, but did appear to offer some protection from 3 to 10 days after treatment. Berry consumption by cedar waxwings did not reduce when blue berries were treated at 32 kg/ha in the laboratory (Cummings *et al.*, 1995). Methyl anthranilate (Bird Shield™) is not found effective for repelling blackbirds from ripening rice and sunflower fields (Werner *et al.*, 2005). Further methyl anthranilate was not found to impart a learned conditional aversion in Canada geese (Belant *et al.*, 1996) and found effective only for short duration and needs repeated application in 3-4 days for measurable repellency to favoured crops (Mason and Clark, 1996). In an experiment carried out at VPKAS, Almora, methyl anthranilate was found not to repel spotted munias, *L. punctulata* damaging the maturing seed crop of radish when sprayed at 1% or kept in scent darts.

Butyramide

Butyramide is reported equally effective as methiocarb in repelling pest birds from seed crops (Schafer, 1978). Butyramide as avian repellent was found to affect the feeding patterns of the house finch, _C. mexicanus_ on broccoli plants when sprayed at 1.7kg a.i/ ha. It is concluded that spot application or application on alternate rows and borders may also significantly reduce the bird damage (Schwab, 1978).

Thiram

Thiram (tatramethyltiuram disulfide), basically a fungicide is also been used for protecting trees, tree seedlings, bulbs and other plant materials from damage by rodents and other mammals (Besser and Welch, 1959; Welch, 1967; Weingartner and Cech, 1974). Witmer and Pipas (1998) reported that thiram at 21% as very effective repellent for porcupines from damaging pine trees whereas the same chemical at 7% is not at all effective. Thiram was found effective as a feeding repellent for white-tailed deer. However the results were reported to be variable and browsing was noticed in treated plots also (Beringer _et al._, 1994).

Ziram

Ziram repelled red-vented bulbuls, _Pycnonotus cafer_ in cage trials and in further field tests, treated plots of dendrobium orchids also received lower levels of bird damage (Cummings _et al._, 1994). Ziram at the recommended rate of 10kg/ha is reported to protect oilseed rape from mute swan, _C. olor_ damage for four weeks (McKay and Parrott, 2002).

II. Plant Chemical Repellents

Capsacin

Repellents based on resin from _Capsicum_ spp. have been used to alter animal behaviour for a variety of species, including bears (Hunt, 1985), ungulates (Andelt _et al._, 1992) and dogs (Bullard, 1985). The resin contains capsaicin, which is the agent that makes them taste hot by stimulating nociceptors of the trigeminal system (Mason _et al._, 1991; Rasmussen, 1994). Technically, capsaicin is 8-methyl-N-vanillyl-6-nonenamide, which irritates skin and mucous membranes. The irritating quality of this stimulation produces a burning sensation that animals find unpleasant. Capsaicin is considered as universal repellent to mammals at concentrations between 10 and 100 ppm (Mason, 1988). It is repellent to Norway rats at 1% concentration (Mason _et al._, 1991).

Capsaicinoids are aromatic amides regarded as strong chemical irritant for most mammals (Rozin _et al._, 1979). Capsicum oleoresin sprays were found success in repelling bears. The mixture of capsicum-resin aerosols, because it affects several sensory systems is used to repel attacking bears and to condition aversively the habituated problem bears (Smith, 1984). Elephants were shown to respond to a capsicum oleoresin aerosol in parks (Osborn, 2002). When 10% capsicum (contains approx. 250,000 Scoville units) was sprayed to intruding elephants it was found much effective and easier technique to repel them from agricultural fields. Elephants generally reacted to the capsicum spray as follows: (1) The first elephant to come into contact with the spray immediately stopped feeding and raised its head in alarm. (2) An audible exhalation of air, then a rumble or roar followed this reaction. (3) The rest of the group froze until the next animal in line inhaled the spray. (4) The elephants then emitted a series of excited trumpets, rumbles and roars followed by a hurried and disoriented exit from the field in the opposite direction of the spray origin. Out of 18 intrusions, 16 successful attempts of repulsion was reported by using capsicum oleoresin whereas only 7 attempts out of 15 were successful by using 4-7 people with several dogs, whips, drums and multiple large fires. The time taken from the start of stimuli up to repelling them is also found very less (2 min) when compared to traditional methods (9.3 min). In no cases did elephants charge after inhaling the spray. Their retreat from the fields was swift (Osborn, 2002). There is a technique of smearing chilli-grease to the fences and found very effective to deter the crop raiding elephants (Hedges and Gunaryadi, 2009) but the disadvantage here is to apply in frequent intervals because it washed off in the rain (Sitati and Walpole, 2006; Graham and Ochieng, 2008). The response of the elephants to aerial spraying of capsicum oleoresin was found more rapid and resulted in prompt withdrawal from the crops without aggression (Osborn, 2002). Capsicum oleoresin has thus far functioned as a viable short term elephant repellent in some areas but does not provide a long term solution to the problem of increasing elephant numbers (Cumming and Jones, 2005).

Capsacin is inoffensive to some birds such as European starlings and rock doves, *Columba livia* (Szolesanyi *et al.*, 1986). Parrots, *Amazona* spp. (Mason and Reidinger, 1983), pigeons (Szolcsanyi *et al.*, 1986), red-winged blackbirds (Mason and Maruniak, 1983), European starlings (Mason and Clark, 1990), cedar waxwings, *B. cedrorum* and house finches, *C. mexicanus* were reportedly found insensitive to capsaicin. But Mason *et al.* (1991) found veratryl acetamide, a structural modified compound of capsaicin is aversive to European starlings. The non irritability of capsaicin to birds is attributed to the lack of *vanilloid receptor subtype 1* (VR1) where the chemical binds to cause action.

Caffeine

Caffeine was identified as a potential avian repellent with relatively low toxicity (LD_{50} = 316 mg/kg) to European starlings, *S. vulgaris* (Schafer *et al.*, 1983). Avery and Cummings (2003) found that 2,500 ppm caffeine reduced rice consumption by 76% among male red-winged blackbirds in captivity. Blackbirds consumed <10% of rice seeds treated with 10,000 ppm caffeine and >80% of untreated rice seeds under field conditions. Thus caffeine may be an effective, economical and environmentally safe chemical repellent for reducing bird damage to newly seeded rice (Avery *et al.*, 2005). Caffeine at 250 ppm reduced rice consumption of blackbirds by 87.2% in choice test and 1000 ppm caused no feeding (Werner *et al.*, 2007).

Cinnamamide

Cinnamamide is a synthetic derivative of the plant secondary compound, cinnamic acid. Laboratory and field trials have shown the compound to be repellent to rock dove (*C. livia*), woodpigeon (*Columba palumbus*), chestnut-capped blackbird (*Agelaius ruficapillus*), rook (*Corvus frugilegus*) and chaffinch (*Fringilla coelebs*) (Crocker and Perry, 1990; Crocker and Reid, 1993; Crocker *et al.*, 1993; Watkins *et al.*, 1995; Gill *et al.*, 1998). Cinnamamide and 3,5 dimethoxycinnamic acid gives an unpleasant taste in the mouth, causing sensations of nausea or as a toxin in the liver or central nervous system. A mixture of cinnamamide to produce a strong initial aversion and 3,5 dimethoxycinnamic acid to maintain it in the longer term are very effective against pigeons (Crocker, 1990).

Cinnamamide was found to be a repellent to rock doves (Crocker *et al.*, 1993; Watkins *et al.*, 1995), Canada geese and house sparrows (Gill *et al.*, 1998). Cinnamamide mixed at a rate of 0.6% a.i/mass of peanut in a suspension concentrate formulation did not allow greenfinches (*C. chloris*), blue tits (*Parus caeruleus*) and great tits (*P. major*) to feed on their favourite food (Gill *et al.*, 1998). Cinnamamide also act as an antifeedant and unpalatable to Norway rats @ 0.25 to 0.5% (Crocker *et al.*, 1993; Spurr and Porter, 1998) and house mice @ 0.1% (Gurney *et al.*, 1996) but not to possums even at a concentration of 0.5% on carrot bait (Spurr and Porter, 1998). Cinnamon oil above 0.5%, reduced the palatability of baits to possums and at 0.3% w/w to rats (Spurr and McGregor, 2003).

Azadirachtin

Azadirachtin, a triterpenoid extracted from neem seeds effectively repels birds especially starlings from particular foods (Mason and Matthew, 1996). Neem products are relatively safe to mammals and humans (Mulla and Su, 1999). The acute oral LD_{50} of azadirachtin is > 5000 mg/kg for Norway rats (Tomlin, 1994). Neem was reported to be a repellent to three species of Kenyan rodents (*Lemniscomys striatus*, *Mastomys natalensis* and *Arvicanthis niloticus*) (Oguge *et al.*, 1997). However, a commercial preparation of neem (0.15% azadirachtin) applied at 2% on carrot baits (at least 0.003% azadirachtin) was not repellent to brushtail possums or laboratory Norway rats in a choice trial (Spurr and McGregor, 2003).

Garlic oil and castor oil

Feeding aversion was observed in European starlings, *S. vulgaris* when fed with garlic oil at 0.1 to 1.0 % (vol./mass). These results are consistent with the possibility that garlic oil may be used to repel these birds (Mason and Linz, 1997). A commercial product containing 65% castor oil (Mole-Med®) was found very effective in repelling eastern moles, *Scalopus aquaticus* from lawns even up to 65 days after treatment (Dudderar *et al.*, 1995).

III. Other Repulsive Compounds

European starlings presented with veratryl alcohol in feeding and drinking trials was found significantly more aversive than vanillin and vanillyl alcohol. The other two chemicals was repelling the bird only at very high concentrations (0.5% mL/mL), which proved lethal (Shah *et al.*, 1991). Copper oxalate is used as seed repellent against bird pests. Quinone is used as a woodpecker repellent on utility poles (Schafer, 1978). Naphthalene is used inside structures as a repellent for pest bird species (Schafer, 1978). Linz *et al.* (2006) observed 81% repellency of red-winged blackbird consumption in rice treated with the label rate of chlorpyriphos (Lorsban®). Fungicide, flutolanil at 20,000 ppm was found to reduce bird damage by 58.8% in broadcasted rice and 50% more seedlings were found in treated compared to untreated fields in seed drilled rice (Werner *et al.*, 2008). The fungicide propiconozole (Tilt®) was reported to reduce the blackbird consumption of rice by 91.5% in laboratory conditions, but did not reduce blackbird consumption of maturing rice under the field conditions (Werner *et al.*, 2008). A reduction in consumption (41%) of sunflower seeds treated with lambda cyhalothrin (Warrior®) by red-winged blackbirds was reported by Linz *et al.* (2006). The chemical toxicant, 4- aminopyridine can be regarded as frightening agent also (DeGrazio *et al.*, 1972). Avitrol is registered for repelling pigeons, house sparrows, blackbirds, grackles, cowbirds, starlings, crows and gulls from feeding, nesting, loafing and roosting sites (Cleary and Dolbeer, 2005).

Siberian pine, *Pinus sibirica* needle oil reduced food consumption by deer mice, *Peromyscus maniculatus* and prairie voles, *Microtus ochrogaster* (Wager-Page *et al.*, 1997). Paradichlorobenzene or napthalene are mentioned as repelling substances for mole control (Marsh, 1962). In warehouses and similar structures where sacked grain is stored, application of powdered sulfur or flake naphthalene scattered over the bags has been found beneficial in reducing rat and mouse damage (Welch, 1967). Quinine hydrochloride and capsaicin discourage coyotes from chewing on irrigation hoses (Werner *et al.*, 1997). Ternent and Garshelis (1999) reported black bears can be deterred using thiabendazole from its favourite food. Polson (1983) used thiabendazole to condition black bears to avoid bee hives.

Copper naphthenate, trimethacarb, zinc naphthenate and ziram are known as mammal repellents. Denatonium benzoate (0.2%) a toxicant, is also reported as a porcupine repellent resulting 25% damage reduction in pine trees, whereas the chemical at 0.065% did not have any repellent action against porcupines (Witmer and Pipas, 1998). Thimet 10G is tried as a repellent for monkey to keep them away from the crops with some success (Chauhan, 1999). Some chemical substances that have provided a certain amount of protection and yet were harmless to plants are lime-sulphur, copper carbonate and asphalt emulsions. Arasan, a fungicide was also used as rabbit repellent (Johnson, 1964). A mixture of lime-sulfur and copper salts when applied to the bark of trees and coniferous seedlings was found to effectively prevent damage by rabbits. These repellents have been used extensively in the forest industry to protect coniferous transplants from hare damage (Welch, 1967). Douglas-fir seedlings were sprayed with TMTD in the nursery bed and found to have excellent protection (Duffield and Eide, 1962). This continues to give protection of deciduous trees from damage by cottontails and jackrabbits also in the field. But attempts to protect haystacks from rabbit depredations through repellents have not met with much success. The treatment is peripheral and the protective barrier provided by the repellent, which is sprayed on the outside surface of the stack, is soon penetrated by the animals, giving them access to untreated hay and rendering the repellent treatment of little value (Welch, 1967).

8.2. Antifertility Agents/ Sterility Agents

It is more practical to prevent animals from being born than to reduce their numbers after they are partially or fully grown and established in a secure environment (Balser, 1964). Fertility control can be achieved by surgical/chemical sterilization, endocrine perturbation sterilization and immuno-contraception. Chemosterilants are widely applied to control the birth rate of troublesome populations of vertebrates as a successful antifertility method. Technologies that induce infertility in wildlife are advancing rapidly, largely due to our increasing understanding of reproductive physiology, as well as the demand for management techniques that reduce fertility rather than increase mortality (Humphrys and Lapidge, 2008). It is a challenging field of research in vertebrate pest management with many opportunities for improvements. Many chemicals act as chemosterilants apart from a few plant compounds and hormones. Anticholesterol compounds were suggested as female gametocytes and expected to cause reduction in egg laying in birds (Nichols and Balloun, 1962; Wetherbee, 1964). Most of the compounds that are proved to be effective oral contraceptive to mammals including tranquilizers, gametocides, antithyroid compounds, hypophyseal inhibitors, insecticides, fungicides and coccidostats (Elder, 1964). Chemicals like triethylene-melamine, α-chlorohydrin, bromocriptine, diethylstilbestrol, mestranol, quinoestrol, ornitrol, progestins, prostaglandins, thioIEPA and metals act as antifertility agents (Bomford, 1990) of both animal and bird pests. Most of the known chemosterilants act on the hypothalamic-pituitary-gonadal axis and affect hormone release or reception (Hygnstrom *et al.*, 1994).

Triethylene-melamine

Triethylene-melamine (TEM) is a classical mammalian/avian chemosterilant. Davis (1959) found that TEM inhibited testicular recrudescence in starlings, causing the testis to become merely an interstitial organ and concluded that the use of a gametocide promises to add sensitivity to control measures. In a field test on red-winged blackbirds, this cytocide inhibitor of meiosis did cause a measurable reduction in hatchability and the number of nestlings produced (Vandenbergh and Davis, 1962). Of great significance is their observation that TEM had no discernible effect on pairing or other mating behaviour. Behavioural changes also did not appear in rats administered TEM (Bock and Jackson, 1957). The difficulty in using TEM is that it has to be made continuously available to the pest througout the breeding season (Howard, 1967).

Alpha-chlorohydrin

Alpha-chlorohydrin is a post-testicular male antifertility agent which is reportedly effective in a wide variety of mammalian species (Kalla, 1976). Though a partial or good effect was reported in Norway rats (*R. norvegicus*), black rats (*R. rattus*) and rice-field rats (*R. argentiventer*), they were found unaffected at acute doses (Kennelly *et al.*, 1970; Bowerman and Brooks, 1971). Different doses of α-chlorohydrin were administered to mature male Polynesian rats, *R. exulans* at 100, 200 and 300 mg per kg body weight to study the potential of antifertility. After 1 day, microscopic lesions were observed in the initial segment of the epididymis of 66.7% rats dosed with 100 mg/kg and in 100% of the 200 mg/kg group. The damage appears to be reversible in animals that survive an acute dose and the drug cannot be considered an effective chemosterilant (Cummins and Wodzicki, 1980). But Indian gerbils, *Tatera indica* males treated with α-chlorohydrin were reported to be infertile with reduced testes and higher sperm mortality (Singla and Prashad, 2009).

Diethylstilbestrol

Diethylstilbestrol (DES), synthetic oestrogen is a female infertility agent. A 5 min contact of nursing mothers of the Levant vole, *Microtus guentheri*, with paper impregnated with DES suspension in soybean oil (125 μg/cm^2) caused sterilization of 60% of the newborn female offspring and one hour contact was sufficient to sterilize more than 90% of them. Treatment of nesting material with DES offers a promising means of reproductive suppression in Levant voles (German, 1985). When silver fox were force-fed with a single dose of 50 mg DES, any time from 9 days before mating to 10 days after mating, failed to produce offspring (Linhart and Enders, 1964). Application of DES in treated oats (0.11% a.i) to female black tailed prairie dogs, *Cynomys ludovicianus* during breeding season resulted in complete curtailment of reproduction (Garrett and Franklin, 1983). The white-tailed deer does treated orally with DES showed a significant reduction in fetal rate but the oral application did not represent the practical method of population control. The intramuscular application of DES was found very effective, but it would only be useful for small population (Harder and Peterle, 1974). Diethylstilbestrol has been commonly used for fertility control in foxes but found to cause temporary sterility (Bomford, 1990).

Mestranol

Mestranol, an estrogen (Howard, 1967) and an antioestrogen compound (Wetherbee, 1964) are regarded as potent oral mammalian antifertility agents (Duncan and Lyster, 1963). Mestranol is demonstrated as an effective sterilizing agent in rats. Both male and female young ones are irreversibly sterilized throughout life when administered to milk feeding adult females. It is found to impair ovulation, fertilization and implantation in adult females if administered orally (Howard, 1967).

Cabergoline and Bromocriptine

Cabergoline and bromocriptine are synthetic dopamine agonists which suppress prolactin secretion from the anterior pituitary in a number of eutherian mammals, including rats, dogs, cats and foxes (DiSalle *et al.*, 1983; Post *et al.*, 1988; Jochle and Jochle, 1993; Onclin *et al.*, 1993; Marks *et al.*, 1996). Cabergoline is reported effective for the management of European badgers and stoats also (Tuyttens and MacDonald, 1998; Norbury, 2000). Oral treatment of beagle bitches during late pregnancy with 5µg cabergoline per kg causes a decrease in prolactin to basal concentrations within 24 h, inducing premature luteolysis and abortion (Post *et al.*, 1988). Oral administration of 20µg/ kg to red foxes, *Vulpes vulpes*, significantly declines prolactin concentration and thus abortion or reduced cub activity. In lactating rats, dogs and cats, both oral and intramuscular treatment with cabergoline inhibits prolactin secretion and effectively reduces or terminates milk production, thereby retarding growth of young. Effect of bromocriptine injection is found success in tammar wallaby (*Macropus eugenii*), Bennett's wallaby (*Macropus rufogriseus*), quokka (*Setonix brachyurus*), Western grey kangaroo (*M. fuliginosus*) and Eastern grey kangaroo (*Macropus giganteus*) (Curlewis *et al.*, 1986; Loudon and Brinklow, 1990; Hinds and Tyndale-Biscoe, 1994). The brushtail possum, *Trichosurus vulpecula* and the fat-tailed dunnart, *Sminthopsis crassicaudata* are reported to be less sensitive to these chemicals (Hearn *et al.*, 1998).

Other chemicals

Thiotepa is basically an insect chemosterilant chemical that has shown a high degree of potential as a male blackbird/starling chemosterilant. 2-amino 5-nitrothiazole, an antifertility agent is potential against many bird species but not on rats (Wetherbee, 1964). Elder (1964) demonstrated 22, 25-diazacholestanol dihydrochloride, an anticholesterol compound as a potent chemosterilant. When this is administered at 0.1% in the diet for 10 days, no eggs were laid by pigeons for 3 months, full fertility among some birds was not reached for 6 months and some remained anovulatory for 12 months. Provera at 0.1% or more in the diet and Arasan at 0.35% inhibited ovulation without severe debilitation of the birds, but the effect was lost as soon as the materials were withdrawn from the diets (Elder, 1964). Treating red-winged blackbirds, *Agelaius phoeniceus* with ornitrol (azacosterol) is a promising measure of management through affecting its fertility. Cracked corn treated with ornitrol, when fed to the birds has reduced the hatching success by 49% compared to the untreated cohort. Ornitrol is used as a chemosterilant for controlling pigeon populations by feeding them with treated whole corn bait (0.1%) for a period, often days and inhibiting their reproduction for about six month (Woulfe, 1968). It is also reported to be effective against Japanese quail, rock doves and house sparrows (Locombe *et al.*, 1986). Mifepristone is reported as the promising chemosterilants for reducing the fertility of European badgers and stoats (Tuyttens and MacDonald, 1998; Norbury, 2000).

Plant compounds and dyes

Gossypol, the phenolic compound found in crude cottonseed oil or meal was found to cause female infertility and also have antispermatogenic effect in rats and monkeys (Qian and Wang, 1984). The oil of the fava olive and cyclopropene fatty acids found in many plants of Malvaceae family such as *Sterculi foetida* suppress hatchability (Howard, 1967). Oleanolic acid isolated from the flowers of *Eugenia jambolana* when administered to albino rats were reported to arrest spermatogenesis for 60 days without causing any abnormality to spermatogenic cells (Rajasekaran *et al.*, 1988). Embelin, extracted from *Embelia ribes* berries possess antiandrogenic activity and alters the testicular histology and gametogenic counts (Agrawal *et al.*, 1986). A black dye known as Sudan Black B when administered to laying adults found to reduce hatchability in birds. In Japanese quail (*C. coturnix*), no egg hatching was observed for 10 days after females were fed as little as 500 mg of Sudan Black B per kg of feed in a single acute dose. When laying females were fed levels lower than 20 mg/kg, the yolks of all eggs laid for about one week were discoloured adversely affecting its hatchability (Wetherbee *et al.*, 1964).

Hormones

It was found that a small concentration of estrogen or a larger amount of androgen can exert an action at the hypothalamic-pituitary areas in new-born rodents in such a way that their future ability to produce ova or sperm and sex hormones is essentially lost (Howard, 1967). Estrogens in the female rat can also cause abortion and in other ways interfere with pregnancy by preventing implantation of embryo (Norbury, 2000). Control of male fertility depends either on the inhibition of spermatogenesis or on methods to render the spermatozoa infertile. Male rats are reported as infertile after 3 months of treatment with methyl testosterone owing to suppression of spermatogenesis (Gao and Short, 1993). Male Brandt's voles, *Lasiopodomys brandtii* when fed with quinestrol 0.006% had effectively affected their reproducing capacity (Wang *et al.*, 2011). Injections of gonadotropin-releasing hormone (GnRH) are reported to be an effective contraceptive in captive wild pigs, rats and rabbits (Stevens, 1986). Results of GnRH injections included reduced ovary and testis weight, reduced levels of testosterone and progesterone and reduced pregnancy rates in pigs (Killian *et al.*, 2003). GonaCon® is a single-shot, multiyear, GnRH immuno-contraceptive vaccine that decreases sexual activity and prevents animals from entering a reproductive state through manipulation of hormones. The GnRH vaccine has been shown to induce contraception in several mammalian species, including rabbits, ground squirrels, domestic cats, domestic and feral pigs and deer (Stevens, 1986; Fagerstone *et al.*, 2008). Oestrogens and progestins prevent ovulation in the female mammal; the frequency of dosing required makes them impractical for use with many species, although they have provided a viable method of contraception for feral cats (Johnson and Tait, 1983).

Immuno-contraception

Egg or sperm antigens may also play a role in immuno-contraception. Infertility can be induced by the blockage of sperm receptor sites on the ovum, altered ovarian follicle growth and function and possibly autoimmune activity associated with the ovary itself rather than using chemosterilants (Hygnstrom *et al.*, 1994). Sperm antigens (Shagli *et al.*, 1990) and zona pellucida peptide (Aitken and Paterson, 1988; Millar *et al.*, 1989) have been shown to cause contraception in rats and in mice. In pest species with a high reproductive potential, the lost output of sterile animals would rapidly be compensated by unaffected, socially-dominant animals. But in this immuno-contraception technique, steroidal functions of the animal's gonads are not meddled and thus, a sterilized animal should be able to maintain its social status and compete for living. Virally vectored immuno-contraception is a novel concept involving inserting foreign genes into live viruses to cause infertility (Tyndale-Biscoe, 1991), which is still in the early stages of research and development. The vaccine deliver by viruses should specifically infect the target pest population and candidate vaccine vectors are identified for mouse, rabbit and fox (Seamark, 2001) and has been shown to work in possums in the laboratory (Cowan, 2000).

There are many drugs and techniques that cause infertility in pest animals and that are both environmentally acceptable and humane. But delivery of fertility control drugs to wild populations and achieving cost-effective animal damage control is the major problem (Bomford and O'Brien, 1992). A significant challenge to the delivery process is providing the international regulators in each jurisdiction with the most relevant data packages they need to assess new products. An essential part of any product registration for free-ranging animals will be the development of species-tailored delivery systems, especially so for non-specific antifertility actives (Humphrys and Lapidge, 2008). More importantly, detailed knowledge is needed on a variety of techniques of application, proper timing, dosage and the dispersal of baits. The results are of greater success when the animals thus made sterile are accepted in its social system and compete with others for food and space.

Many scientists have postulated that a given number of sterile individuals in a population exert a much greater biological control pressure on that population than removal of that same number of fertile individuals. Applying antifertility agents to species having one litter per year is much simpler than application to those which produce several litters a year. The latter may require drugs that produce permanent sterility to make baiting practical; otherwise bait would have to be repeatedly applied or continuously made available. Fertility control effectively reduced density in rice field rats if a minimum of 50% of females were sterilized (Jacob *et al.*, 2004) and sterilizing such a large population is a difficult process. If achieved, the effect of sterilization is carried out generations to generations leading to substantial decline in rodent population (Gao and Short, 1993).

Further, chemosterilants cannot change the general equilibrium position of pest populations permanently, so repeated applications are required to restrain the populations from again increasing beyond the economic threshold density (Howard, 1967). Moreover, many of the chemosterilants are highly teratogenic, mutagenic and/or carcinogenic, so it should be used with utmost care (Schafer, 1978).

8.3. Tranquilizers and Immobilizing Agents

Tranquilizers are used to temporarily sedate an animal so that it may be handled or captured safely. Darts filled with tranquilizers are shot through a tranquilizer gun on animals to get them anesthetized. A tranquilizer can be a sedative, an anesthetic or a paralytic agent. Tranquilizing agents are classified in to two *viz.*, neuromuscular blocking agents and central nervous system (CNS) depressants. Neuromuscular blocking agents have a relatively rapid onset and short duration of action but they do not possess sedative, analgesic or anesthetic properties. Central nervous system depressants do produce desirable sedative, analgesic and anesthetic effects and a combination of CNS depressants results in more desirable immobilization characteristics. Because of many disadvantages, neuromuscular blockers are not used as much as central nervous system depressants for anesthetizing animals. Mortality rates from the use of neuromuscular blocking agents range from 10 to 20% whereas the mortality rate with the centrally acting drugs is less than 3%. Some of the important tranquilizing agents are discussed here under:

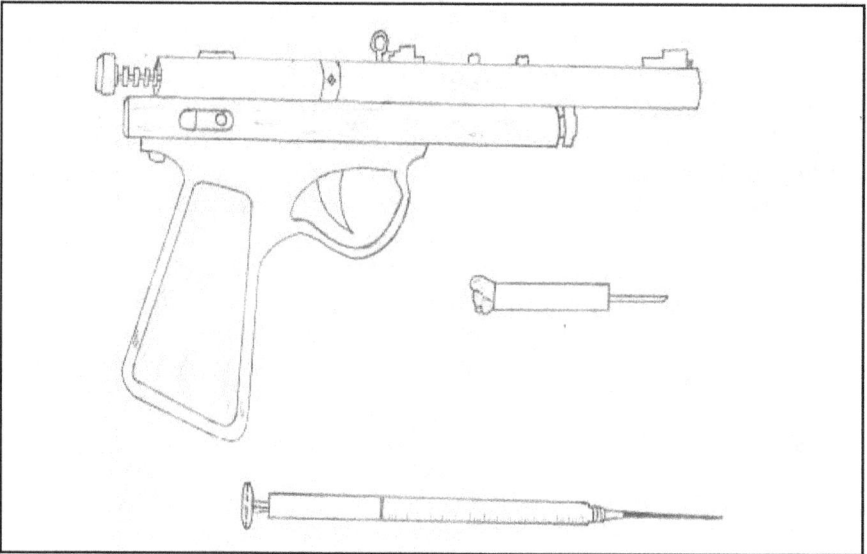

Figure 8.1: Tranquilizer gun and darts

d-Tubocurarine: Centuries before, Indians of South America used poisoned arrows with curare to kill animals for food. Curare is extracted from poisonous plants having the active ingredient of d-tubocurarine. It produces complete muscle paralysis (Koelle, 1975) but not have any significant effect on central nervous system (Savarie, 1976).

Succinylcholine: It is an important neuromuscular blocking agent of the depolarizing type and widely used for the capture of wild animals with an advantage of a rapid onset of action. However, there is no effective antidote and there is a narrow safety margin between doses that produce immobilization and those that produce respiratory paralysis (Savarie, 1976).

Nicotine: Basically nicotine is a ganglionic stimulating agent (Volle and Koelle, 1975), but in the doses employed for the capture of wild animals, it causes neuromuscular blockade causing muscular paralysis. Animals became immobilized in 2.5 to 15 min after receiving 200-300 mg nicotine salicylate intramuscularly (Crockford *et al.*, 1957) and the recovery time ranged from 15 to 60 min.

Barbiturates: A wide range of desirable applications including sedation, hypnosis, anticonvulsant activity and anesthesia can be obtained from chemicals known as barbiturates. Barbiturates can also be considered as intramuscular immobilizing agents but with prolonged induction times. Thiopental, hexobarbital and pento-barbital produced anesthesia within 5 min when administered intravenously to white-tailed deer (Severinghous, 1950; Piperno, 1965). These are also used to anesthetize black bear, *U. americanus* (Erickson, 1957).

Ether and Chloroform: Black bear caught in traps are anesthetized by applying ether or chloroform to confine them in the trap (Erickson, 1957; Black *et al.*, 1959). Ether acts in the neuromuscular junction and can also potentiate neuromuscular blocking agents. Marked salivation occurs during the induction and recovery phases of ether anesthesia.

Xylazine: It produces sedation, analgesia and muscular relaxation. In majority of mammals, xylazine have sedative and muscle relaxant properties accompanied by good analgesia and immobilization with minimal excitement. The advantage of xylazine is high safety margins *i.e.*, 8 fold in white-tailed deer (Roughton, 1975). Ketamine-xylazine is an effective anesthesia widely used for many animals especially rabbits (Popilskis *et al.*, 1991).

Phencyclidine: It is reported to cause excitatory symptoms in mice and rats whereas sedation in cats, dogs and monkeys. Harthoorn (1962, 1963) has used phencyclidine in combination with morphine or diethylthiambutene and scopolamine. This combination produced a large increase in the margin of safety and better immobilization characteristics in elephant, rhinoceros, hippopotamus, giraffe and buffalo.

Thiambutene: It produces mild sedation, analgesia and narcosis (Owen, 1955). Thiambutene has only limited application in capturing wild animals because of the large amount required to be effective and slow onset of action. A combination of thiambutene, phencyclidine and scopolamine has been used to anesthetize topi antelope, *Damaliscus korrigum* and hippopotami, *Hippopotamus amphibus*.

Ketamine: Ketamine has a faster onset, shorter duration of action and a lower incidence of convulsing activity (Chen *et al.*, 1966). The wide margin of safety and relatively short recovery period associated with ketamine make it a useful drug for the capture of carnivores especially spotted hyaenas, *Crocuta crocuta* (Smuts, 1973). A dose of 2.5 mg/kg of ketamine hydrochloride induced recumbency in 5 min to free ranging red deer, *Cervus elaphus* when fired with plastic projectile syringes from a dart gun. The distance covered by the animals between darting and recumbency ranged from 40-300 m (Arnemo *et al.*, 1994).

Diazepam: It is regarded as a good tranquilizer because of its long lasting action, wider safety range and do not affect much on the respiratory centre. A diazepam tablet of 500mg and 1g are found effective for 24 to 48 hrs for fox, *Vulpes* sp. and coyote, *Canis latrans*, respectively (Balser, 1965).

Carfentanil: Captive mule deer were reported successfully immobilized with carfentanil at a dose of 0.2 to 0.5 mg/kg with an induction time of 3 to 12 min (Jessup *et al.*, 1964).

Propiopromazine hydrochloride: A tranquilizer trap device with propiopromazine hydrochloride fitted in foot hold trap is used successfully in trapping grey wolves, *C. lupus* with very less injuries (Sahr and Knowlton, 2000). Propiopromazine hydrochloride and ascorbic acid, an antioxidant tested showed 67%, 90% and 67% tranquilization in feral dogs, coyotes and wolves, respectively (Savarie *et al.*, 2004).

Tiletamine-zolazepam: It is found effective against river otters, *Lutra canadensis* at a dose of 4 mg/kg. Anesthetic induction was reported rapid and smooth but with prolonged recovery (89 min) (Spelman, 1997). Tiletamine-zolazepam-xylazine is found very effective on rabbits as an anesthesia (Popilskis *et al.*, 1991). The use of tiletamine-ketamine-xylazine is very effective on feral cats with very less mortality (Lindsay *et al.*, 2002). zolazepam-tiletamine at 0.9–2.8 mg/kg was successfully used for anesthetizing free ranging brown bears, *Ursus arctos* (Evans *et al.*, 2012).

Morphine and Its Derivatives: Morphine is a natural chemical obtained from opium poppy plant, *Papaver somniferum*. The immobilizing characteristics of these chemicals include stupor or insensibility, loss of fear and anxiety and marked analgesia. Frequently, animals stand with heads drooped and if they fall to the ground they lie in sternal recumbency. These chemicals have a wide safety margin with minimal side effects. Effective antagonists are available which quickly reverse the immobilization effects. Etorphine, a morphine derivative with 1000 times effective than morphine is being used extensively for capturing animals (Harthoorn and Bligh, 1965; Wallach, 1966; Woolf, 1970; Alford *et al.*, 1974; Roussel and Patenaude, 1975). Etorphine-xylazine mixture having wide safety margin is reported as appropriate drug to capture Malayan gaur, *Bos gaurus hubbacki* (Conry, 1986).

Atropine and Scopolamine: Both atropine and scopolamine are anticholinergic drugs and are used frequently to antagonize side effects of salivation and cardiovascular disturbances such as decreased heart rate and low blood pressure.

Phenothiazines: Different phenothiazine derivatives such as chlorpromazine, promazine and acetylpromazine are used but are not effective immobilizing agents because of the slow onset of action and prolonged recovery periods. They are best used in combination with other CNS depressants to produce potentiation of more desirable immobilizing characteristics. They are effective sedatives for use in transportation of animals.

Application in Pest Management

Animals are anesthetized by using 1cc syringes modified into tranquilizer darts and shot using a blow gun to capture them (Brockelman and Kobayashi, 1971). Black bears, *U. americanus* involved in livestock depredations were effectively anesthetized by injecting tiletamine/ zolazepam using a dart gun or jab stick and relocated to ranges with no domestic sheep rearing (Armistead *et al.*, 1994).

Tranquilizer tabs attached to traps are some form of anesthetizing the trapped animal (Balser, 1965; Linhart *et al.*, 1981) to reduce struggling and secondary injuries to them and also reduces the chance of escape (Zemlicka and Bruce, 1991). Foot hold traps were evaluated for injuries caused for the trapped wolves, showed only less and minor leg injuries with traps fitted with a tranquilizer device, when compared to trap without the device which caused severe injuries to 57% of the trapped animals (Zemlicka *et al.*, 1997). Diazepam (Balser, 1965), propiopromazine hydrochloride (Linhart *et al.*, 1981; Zemlicka and Bruce, 1991), ketamine hydrochloride and xylazine (Chamberlain *et al.*, 1981) are reported as effective tranquilizing agents for trap use. Non-targets caught in foot hold traps equipped with tranquilizer trap devices with propiopromazine did not found to die due to drug overdoses and injuries were found significantly less than traps without tranquilizer devices (Sahr and Knowlton, 2000).

Immobilizing Agents for Birds

There are chemical compounds which are used to immobilize pest birds and thus can be captured easily. In feeding studies with penned cowbirds, *M. ater* and coturnix quail, *C. coturnix*, chlordiazepoxide and sodium pentobarbital were found to cause narcosis (Peek, 1966). Tribromoethanol and Alpha-chloralose were successfully used for the capture of Canada geese, *B. canadensis* (Crider and McDaniel, 1967) and wild turkeys, *M. gallopava* (Evans *et al.*, 1975).

Alpha-chloralose is an immobilizing agent for use in capturing waterfowl, coots and pigeons. Most effective dose (MED) is reported to be 30 and 60 mg/kg of body weight for capturing waterfowl and pigeons, respectively. A total of 587 waterfowls and 1,370 pigeons were reported to be captured with 8% mortality for ducks, 0% for geese and 6% for pigeons at these respective doses (Woronecki *et al.*, 1992). Baits treated with alpha-chloralose at 30 mg/kg dosage resulted in 50% capture success of herring gulls, *L. argentatus* and with no mortality. Alpha-chloralose at 60, 120 and 180 mg/kg showed its first effect on pigeons, *C. livia* from 153, 53 and 33 min after feeding, respectively (Belant and Seamans, 1999). Incorporation of alpha-chloralose into bread baits is ideal for selectively capturing ducks, geese and coots that can be hand-fed at urban ponds and parks. Corn baits are recommended for pigeons or groups of waterfowl or coots that cannot be individually baited. Birds ingesting a clinical dose of alpha-chloralose can be captured in 30 to 90 min. Complete recovery normally occurs within 8 hrs but can take up to 24 hrs (Cleary and Dolbeer, 2005).

Methiocarb has also been successfully used on baits to capture large numbers of blackbirds for banding and laboratory study purposes. It has also been tested on pigeons and waterfowl; however, its efficacy on these species is marginal. Tribromoethanol is an efficient immobilizer of pigeons and some wild game birds. Methoxymol combines many of the positive aspects of tribromoethanol and alpha-chloralose, but acts more quickly and is less lasting. It has successfully been used on a large number of species, often with dramatic results (Schafer, 1978). Other immobilizers in general use are barbiturates, hypnotics, etc. but their cost and availability makes them less useful. Immobilizers or tranquilizers form the important part in situations where the animal or bird pest is not needed to be killed but captured for laboratory purposes or for other tagging studies or for relocation programmes. Even some immobilizers like alpha-chloralose are reported as more effective than chemical toxicant in removing herring gulls (Seamans and Belant, 1999). Tranquilizers are especially used to capture big mammals and to relocate them or put them back in to thick forests away from human habitats.

8.4. Chemical Toxicants

Many different measures are available for the management of pest birds and animals which are discussed in the other chapters. If the pest problem is so severe and become difficult to control by any other means then we may sought for chemical toxicants. Many a times, chemical toxicants are regarded as quicker and easier way to solve the pest problems but proper care should be taken for it residues, toxicity to non-targets and other environmental problems. Researchers have demonstrated that strychnine treated grain for the control of English sparrows, cyanide for pigeons and phosphorus treated eggs for carrion crows as the efficient control measures (Seubert, 1964). Later 4-aminopyridine was introduced as a bird toxicant /repellent and extensively used. Some of the important toxicants used for bird control are described below.

4- aminopyridine

Though 4-aminopyridine (4AP) is regarded as a chemical frightening agent for the protection of crops from birds (Besser, 1978), its action as toxicant to cause mortality is well established. It is toxic to most avian and mammalian species, with acute LD_{50}s generally below 10 mg/kg (Schafer *et al.*, 1973). DeGrazio *et al.* (1972) and Stickley *et al.* (1972) showed that 4-aminopyridine baits can be used effectively to protect ripening field corn from blackbird damage. Minimum dilution ratios of 4-aminopyridine that produced more than 50% mortality were 1:29 for adult bobwhites, *Colinus virginianus* and 1:99 for mourning doves, *Zenaida macroura*. Median lethal concentrations (LC_{50}) of 4-aminopyridine to mourning doves and coturnix quail, *Coturnix coturnix* fed exclusively on uniformly treated laboratory rations for 28 to 40 days were found to be 316 and 479 ppm, respectively. Mortality in all avian species appeared to be the result of acute poisoning and no evidence of cumulative toxicity was noted (Schafer and Marking, 1975).

Application of 1.25% aminopyridine on cracked corn reduced red-winged blackbird numbers at a feedlot by 95% within 2 hrs after the bait was spread and only fewer birds visited this lot for the next 3 week period. Aminopyridine 1% cracked corn bait reduced redwings for the entire damage period (from November to March) with four applications. After each application redwing populations fell sharply within a few days of baiting and then began a gradual increase toward the original number. Birds other than blackbirds like house sparrows (*Passer domesticus*), starlings (*S. vulgaris*), common grackles (*Quiscalus quiscula*), yellow-headed blackbirds (*Xanthocephalus xanthocephalus*), black-billed magpies (*Picapica hudsonia*), white-crowned sparrows (*Zonotrichia leucophrys*) and common pigeons (*C. livia*) were also controlled (Woronecki *et al.*, 1967). In corn field, 4% aminopyridine sprays reduced blackbird damage by 78%. Pheasants, *Phasianus colchicus* and mourning doves, *Z. macroura* were not affected by this spray apart from many song birds (Woronecki *et al.*, 1967). It is found excellent in protecting sweet corn and sprouting rice against blackbird damage (Besser, 1976). Mott *et al.* (1977) reported that the brown-headed cowbirds, *M. ater* did not leave treated fields in response to affected birds and caused damage to grain sorghum. Sprays of 4- aminopyridine @ 4 to 16% directed on ears of partially-husked corn and heads of ripening sunflowers to protect them from damage by monk parakeets, *M. monachus* (Calvi *et al.*, 1977).

Aminopyridine millet baits have been tried for protecting ripening millet from red-billed quelea, *Q. quelea* (DeGrazio *et al.*, 1972) and for protecting ripening rice from quelea and other weavers. This is recommended at the rate of 1:99 dilutions for baiting but a lesser dilution of 1:33 was found good in reducing the crop damage (Besser, 1978). Another method of broadcasting the bait thinly by hand over an area covering approximately 25% of each field in swaths 8 to 10 rows wide and about 30 to 35 rows apart was found to reduce blackbird by 85%. Killing of a few blackbirds or only a few affected birds in a field is enough to frighten away others in the flock (Woronecki *et al.*, 1967). In very low doses/ sublethal doses, aminopyridine cause disorientation in birds and often results in erratic behaviour, distress calling, startling other birds and causing them to fly away (Timm, 1994a). To frighten the birds and not to kill, only a portion of a bait presentation (e.g., 10% of corn kernels) is treated with the chemical so that only a small number of the birds are affected but all will get dispersed.

3-chloro-4-methyl benzenamine

3-chloro-4-methyl benzenamine is a slow-acting avicide popularly known as DRC1339 and marketed as Starlicide® for the control of blackbirds, starlings, pigeons, crows, ravens, magpies and gulls. DRC-1339 is highly toxic to most sensitive bird species (LD_{50}'s range from 1 to 10 mg/kg), allowing a toxic dose to be placed on a single bait. Many of the predatory birds except owls and most of the mammals except cats are non sensitive to the chemical. The mode of action of the chemical in sensitive birds is irreversible kidney and heart damage and death normally occurs 1 to 3 days following ingestion. DRC-1339 is metabolized and excreted from all animals very quickly, with 90% or more of the compound lost within 2 hrs. Thus residues generally range from 0 to less than 0.1 ppm when death occurs, reducing the chances of secondary poisoning.

3-chloro *p*-toluidine hydrochloride

3-chloro p-toluidine hydrochloride (0.1%) marketed as Starlicide Complete® is used for controlling starlings and blackbirds. It is toxic to other types of birds in differing amounts, but will not kill house or English sparrows at this level. Mammals are generally resistant to the toxic effects. Poisoned birds experience a slow, non-violent death. They usually die 24 to 36 hrs after feeding, often at their roost. Generally, few dead starlings will be found at the baiting site. Poisoned starlings are not found dangerous to scavengers or predators (Johnson and Glahn, 1998).

Toxicants for Mammals

The primates *i.e.*, baboons, *Papio ursinus* and monkeys, *Cercopithecus aethiops* were thought to be controlled by organized shooting and trapping but not found effective. But baiting with poison baits (thallium) in maize or fruits was found very effective but to be done with utmost care (Hey, 1964). Poisoning has the potential to greatly reduce the cost of controlling stoats compared with traditional methods of trapping because fewer visits to the control area are required and especially because some animals are often difficult to trap (King, 1989). Coblentz and Baber (1987) reported that poisoning was 11 times cheaper than shooting and 80 times cheaper than trapping especially for a pig eradication program. Use of strychnine, arsenic and cyanide date back hundreds if not thousands of years and are considered as ancient toxins. Much innovations and proliferations occur during 1950-1980 during which many toxins like zinc phosphide, warfarin, cholecalciferol, pindone, coumatetralyl, bromodiolone, brodifacoum, flocoumafen etc were developed. Later, animal control relied mostly on anticoagulants and the recent development and use of para-aminopropiophenone is also evident. Animal toxicants are classified as acute toxicants, anticoagulants, fumigants and others. The detailed description on toxicants other than fumigants is given here under.

1. Acute toxicants

(1) Para-aminopropiophenone (PAPP)

Para-aminopropiophenone kills animal by preventing red blood cells from carrying oxygen causing methaemoglobinaemia, resulting in central nervous system anoxia. PAPP is developed as a toxin for feral cats, weasels, red foxes, wild dogs and stoats (Shapiro *et al.*, 2010; Dilks *et al.*, 2011) but not effective against wallabies and possums (Fisher *et al.*, 2008). The oral LD_{50} values of feral cat, dog, coyote, stoat, bobcat and ferret were reported to be 5.6, 7.5, 5.6, 9.3, 10.0 and 15.5 mg/kg, respectively (Murphy *et al.*, 2011). Carnivores appear to be much more susceptible to PAPP than birds, so it potentially has high target specificity. In pen trials, PAPP 20 to 34 mg/kg was lethal for feral cats and 37 to 95 mg/kg for stoats (Eason *et al.*, 2010b). Stoat abundance was reduced by 87% when 13 mg PAPP poisoned meat baits were given to the animal, thus found much effective (Dilks *et al.*, 2011). PAPP which is a humane, causing low risk of secondary poisoning, low toxic to birds, availability of antidote and biodegradable, forms an ideal toxin for animal management.

(2) Sodium monofluroacetate (1080)

Sodium monofluoroacetate popularly known as 1080 is the widely used toxicant used for the management of mammal pests. It is an acute toxin and death of stoats, cats, rats and possums occurred very quickly after poison was deployed. Aerial poisoning of 1080 baits for possum control not only controlled the target pest but also reduces the ship rat population by over 90% (Innes *et al.*, 1995). It is also reported very effective against all rodents. Ground squirrels, *Citellus beecheyi* and *C. oregonus* are reported to be controlled by use of 1080 four to six weeks during the breeding season (Dana, 1962). Dama wallabies are effectively controlled by using 1080 in New Zealand (Warburton, 1990). Acute oral LD_{50} of 1080 poison for feral pigs, *S. scrofa*, obtained by moving average and probit analysis methods are 1.04 and 1.00 mg/kg, respectively. These values are slightly higher than LD_{50s} obtained for pigs by intraperitoneal dosing but similar to those obtained by oral dosing for other eutherian mammals. Signs of poisoning, either vomiting or increasing lethargy or laboured respiration, appeared in 6.2 h and deaths in >16 h after dosing (McIlroy, 1983). Pigout®, bait containing 1080 is used in Australia to achieve population reduction of wild pigs of at least 73% (Cowled *et al.*, 2006). Dried 1080 meat baits have been shown to be very effective for fox control at 0.15 mg/kg of bodyweight (McIlroy and King, 1990). Quarterly aerial baiting with 1080 at a density of 10 baits/ km^2 successfully controlled red foxes over a 12 month period (Moseby and Hill, 2011). Successful campaigns for removal of feral cats were done using 1080 as a primary poison (Campbell *et al.*, 2011).

Though reported as very effective, it is a broad spectrum toxin so secondary toxicity is a possible threat for successful pest management programmes. Several studies have already demonstrated secondary poisoning of predators when using 1080 (Gillies and Pierce, 1999; Murphy *et al.*, 1999). Gillies and Pierce (1999) found that radio-tagged small mammalian carnivores died of secondary poisoning during 1080 bait station poisoning operations. Similarly, all the radio-tagged stoats died shortly after aerial application of 1080 for rodent and possum control (Murphy *et al.*, 1999). When these secondary predators are also considered as pests, then poisoning is found effective. But large proportions of marked tomtits, *Petroica macrocephala* and New Zealand robins, *Petroica australis* were poisoned (primary poison) by aerial application of 1080 on carrot baits in New Zealand (Powlesland, 1998; Powlesland *et al.*, 1999). A total of 36 species out of a selection of 40 non-target species likely to feed on poisoned baits kept for pigs are more susceptible to 1080 than pigs. In practice many other factors such as bait acceptance will govern what proportions of target and non-target populations will be poisoned. Alterations on methods of poisoning or baiting techniques could minimize the risk non-target animals (McIlroy, 1983). Pigout® is designed to attract only wild pigs and not other native wildlife species (Cowled *et al.*, 2006). There are also reports on the influence of temperature on the toxicity of 1080 to mice, guinea pigs, brushtail possums and raccoons (Oliver and King, 1983; Eastland and Beasom, 1986). In USA, 1080 use is restricted to a livestock-protection to protect sheep and goats from coyotes, *C. latrans* (Fagerstone *et al.*, 1994).

(3) Cyanide

Sodium cyanide poison is reported potentially a more rapid and humane method to control wild dogs and wallabies than 1080 poison (Hooke *et al.*, 2006; Eason *et al.*, 2010a). The lethal dose (LD$_{50}$) for cyanide on possums is 8.7 mg/kg (Thomas and Ross, 2007). After the ingestion of cyanide, the average time to onset of ataxia in possums was found to be 3 min, loss of consciousness was 6.5 min and the time to cessation of breathing was 18 min. Thus cyanide is a rapid-acting toxin with few undesirable signs from the welfare perspective (Gregory *et al.*, 1998). Cage trials carried out on Bennett's wallabies, *M. rufogriseus* confirmed the potential of cyanide for wallaby control (Eason *et al.*, 2011). More than 90% mortality of radio-collared tammar wallabies fed with cyanide pellets that remained at the bait site was reported.

In a 30 ha field trial, with cyanide pellets (Feratox®) @ 200 mg bait to Bennett's wallabies, 110 dead wallabies were found less than 50 m from the baiting sites which confirms that Feratox® as an effective ground-based alternative to 1080 for wallaby control (Ross *et al.*, 2011). Feratox® was reported very effective against possums when kept in paper bags located on trees 20 m apart and used with flour and icing sugar blaze (Thomas *et al.*, 2003). Cyanide pellets have a low secondary poisoning risk and proven humaneness in animal control (Gregory *et al.*, 1998) but sublethal doses are reported to cause bait aversion (Morgan *et al.*, 2001).

(4) Cholecalciferol

A form of Vitamin D, calciferol/ ergocalciferol interferes with calcium homeostasis, causing the mobilization of calcium from the bone matrix and increased uptake in the gut (Meehan, 1984; Timm, 1994b). Victims usually die from hypercalcaemia, kidney failure and/or the side-effects of soft-tissue calcification, particularly metastatic calcification of the blood vessels and nephrocalcinosis (Meehan, 1984). Calciferol or cholecalciferol (Vitamin D_3) is an acute poison and can be readily formulated as onetime feed bait requiring no pre-baiting (Eason and Wickstrom, 2001). It is less suitable for outside use because of degradation problem and accidental poisoning of non-targets are difficult to treat (Mason and Littin, 2003). Calciferol brings low secondary poisoning risks, as rodents tend to cease eating after consuming relatively small amounts; it is also quickly metabolized within the rodent's body (Stone *et al.*, 1999; Eason and Wickstrom, 2001).

(5) Zinc Phosphide

It is a commonly used rodenticide, acts by producing phosphine gas in the stomach which kills after a single dose. It is useful in rapid population reduction programmes but has a problem of bait shyness and thus requires pre baiting (Meehan, 1984). The persistence of the compound in the gut of the poisoned animals for several days pose a threat of secondary poisoning (Guale *et al.*, 1994; Timm, 1994b). However, it does not accumulate in their muscles or other tissues (Timm, 1994b; Sterner and Mauldin, 1995) nor within the predators themselves, so the risk is low. Thus, as long as any single dose eaten is not too great, predators will experience no ill effects even if fed on poisoned rodents over several days (Meehan, 1984).

2. Anticoagulants

Anticoagulant poisoning is by far the most common means of rodent removal, being the basis of about 95% of rat and mouse control in the USA (Timm, 1994b) and 92% in UK (McDonald and Harris, 2000). They act by interfering with Vitamin K-1 metabolism. Warfarin was the first important anticoagulant, but because of the genetic resistance (Quy *et al.*, 1992; Smith *et al.*, 1994) second generation compounds such as brodifacoum, difenacoum, bromadiolone etc are being evolved (Mason and Littin, 2003). Second generation anticoagulants have increased the feasibility of rodent eradications (Howald *et al.*, 2007). Anticoagulants are extremely effective, cheap and easy to use, although some require repeated baiting and thus labour-intensive (Meehan, 1984) and if used correctly, they cause little bait shyness (Proctor, 1994). Furthermore, they are relatively safe if accidental poisoning occurs: their slowness of action allows several days for medical intervention by administering Vitamin K-1 and blood products (Padgett *et al.*, 1998; Sheafor and Couto, 1999). Though secondary poisoning of anticoagulants is minimal it cannot be ignored. Fatal secondary anticoagulant poisoning has been implicated in the deaths of red foxes, owls, buzzards, kites, corvids and many other dogs and cats (Newton *et al.*, 1990; Martin *et al.*, 1994; Padgett *et al.*, 1998; Shore *et al.*, 1999; Stephenson *et al.*, 1999).

(1) Brodifacoum

Brodifacoum is a potent second-generation anticoagulant rodenticide used worldwide for rodent control (Eason *et al.*, 2011) very effective especially in controlling rat populations (Innes *et al.*, 1995). Difenacoum, an analogue of brodifacoum killed 82% of mice population in 31 days toxicity tests and also found to control 96% house mice, *Mus musculus* in field trials (Rowe *et al.*, 1981). Possums are susceptible to brodifacoum, but not to first-generation anticoagulants, hence brodifacoum has been the only alternative to 1080 that effectively targets both possums and rodents (Eason *et al.*, 2011). Though effective, secondary poisoning was reported to kill stoats following brodifacoum poisoning operations that killed rodents and possums (Alterio *et al.*, 1997; Brown *et al.*, 1998). Similarly, a large proportion of marked robins were poisoned when cereal pellet baits containing brodifacoum were hand-broadcast in a beech forest (Brown, 1997). Secondary poisoning of cats leveraged through rodents, poisoned with brodifacoum is also reported (Campbell *et al.*, 2011).

2. Flocoumafen

The anticoagulant rodenticide, flocoumafen is reported equally effective as difenacoum, bromadiolone and brodifacoum but with an advantage of quicker kill of mice, *Mus musculus*. Full mortality of free-living and laboratory reared populations of house mice, *M. musculus* was obtained with flocoumafen at 0.005% within 10 days. It gave 87-100% mortality in field trials also (Rowe *et al.*, 1985). Flocoumafen 0·005% was found to completely control warfarin-resistant Norway rats infesting farm buildings in 10-21 days (Buckle, 1986). Large-scale control of rats and possums by aerial or bait station poisoning operations in several forests was attempted using flocoumafen (Henderson *et al.*, 1994; Murphy *et al.*, 1998; Spurr and McGregor, 2003).

(3) Bromadiolone

Bromadiolone, a hydroxy coumarin is developed mainly as an anticoagulant for the management of warfarin resistant rodents. It is regarded as an indirect anticoagulant since it interferes with Vitamin K activity and has 10 times greater potency than that of warfarin in rabbits (Meehan, 1978). It is effective against wide range of rodents including gophers, squirrels, mice and voles (Marsh, 1977). The acute oral LD_{50} of rat and mice are found to be 0.65 and 0.99 mg/kg, respectively (Meehan, 1978). Bromadiolone 0·005% completely killed *R. norvegicus* and *R. rattus* populations not resistant to warfarin after exposure to the poison for 1 and 5 days, respectively whereas warfarin resistant *R. norvegicus* were all killed in 4 days and resistant *M. musculus* in 12 days (Redfern and Gill, 1980). Over 90% of rodents were found dead when baited with 0.005% bromadiolone in field trials tested for seven days (Meehan, 1978; Rowe *et al.*, 1981). House mouse, *Mus domesticus* was reported effectively controlled by using bromadiolone in soybean crop (Singleton *et al.*, 1991). A single application of 0.005% bromadiolone, placed in bait stations (200 g/station) 20 m apart throughout a mouse infested soybean field, effectively reduced the pest by 43-62% with respect to control. But the overall benefit from baiting was found marginal, because mouse population after 42 days of baiting was still 43% of pre-baiting levels. Few mice which survived baiting and mice from other fields apparently replaced those killed (Twigg *et al.*, 1991). Predators, mainly foxes, *V. vulpes* and buzzards, *Buteo buteo* were potentially exposed to anticoagulant compound, bromadiolone via contaminated prey in some instances but do not found to have much risk (Berny *et al.*, 1997). The maximum tolerable single oral dose of bromadiolone to dogs is 10 mg/kg and to cats 25 mg/kg, thus have a higher safety ratio (Meehan, 1978).

(4) Coumatetralyl

Coumatetralyl bait (0.0375%) kept either near the dens or at entry points along the fence line of a flower garden was found very effective in controlling porcupines, *Hystrix indica*. Bait consumption increased up to 7th day and then it steadily decreased reaching zero level on the 15th day, after that there was no porcupine activity indicating 100% reduction of porcupine population (Khan, 2008). Warfarin-resistant rats, *R. norvegicus* were significantly less susceptible to coumatetralyl than non-resistant rats. It is considered that coumatetralyl 0·05% bait is a good alternative to warfarin against non-resistant rats. Continuous use of coumatetralyl might eventually produce an increase in the incidence of resistance to both anticoagulants (Greaves and Ayres, 1969). Cross resistance was reported in coumatetralyl resistant population of Norway rats to bromadiolone also (Endepols *et al.*, 2007).

(5) Diphacinone

Diphacinone is a potent hypo-prothrombinemic agent used as an oral anticoagulant rodenticide. The reported acute oral LD_{50} was 1.9 mg/kg for rats and 5-15 mg/kg for dogs and cats. Consumption of 0.01% of diphacinone for 3 days was fatal to 80 % of deer mice, *Peromyscus maniculatus*. In field tests, 0.01% diphacinone bait was broadcasted at 1kg /acre and found very effective (Howard *et al.*, 1970). An island-wide grid of elevated bait stations containing 0.005% diphacinone bait blocks were used to eradicate the rats, *R. rattus* from Buck Island Reef National Monument in the Caribbean Sea during 1998–2000. No non-target losses resulting from the baiting program were observed (Witmer *et al.*, 2007). The LD_{50} of diphacinone to coyotes was found to be 0.6 mg/kg and death lasts in ten days. Secondary poisoning is noticed in animals that repeatedly feed on tissues containing not less than 0.5 ppm diphacinone but no risk was observed in golden eagles, *Aquila chrysaetos* (Savarie *et al.*, 1979).

(6) Pindone

Pindone has many advantages includes, that the animals feel no ill-effects prior to consuming a lethal dose, thereby preventing learned bait or toxin aversion and so no need of pre-baiting. It was found effective for the control of rabbit populations in Australia (Oliver *et al.*, 1982). Field trials using 150 and 250 ppm pindone baits achieved a mean kill of 94 and 97% rabbits, respectively (Nelson and Hickling, 1994). Poisoning with pindone reduces the target rabbit population and mice, *M. musculus* and thus an integrated management of both with one methodology (Parkes, 1996). Pindone is much less likely to cause persistent residues in livestock than is bromodialone or flucoumafen, although it is somewhat more persistent than 1080 (Eason *et al.*, 1993). Pindone may be secondarily hazardous to scavengers such as the harrier hawk, *Circus approximans* (Calvin and Jackson, 1991).

(7) Warfarin

It is an anticoagulant widely used earlier as a rodent toxicant, has been used to control and nearly eliminate wild pig populations in Australia (Saunders *et al.*, 1990). Experimental studies by Buckle *et al.* (1984) described field tests of warfarin for control of rice field rat, *R. argentiventer*. But in 1958, a population of *R. norvegicus*, in Scotland, became resistant to the most commonly used anticoagulant, warfarin (Boyle, 1960).

The important chemicals used for mammal management, their effective dose, persistence and killing time are given in the 8.1.

Table 8.1: Chemicals used for mammal management: Their effective dose, presistence and killing time

Compound	Efficacy		Residue	
	Dosage	**Mean time for death**	**Half-life Value**	**Persistence of residues***
Para-aminopropiophenone	226 mg	43 min in red foxes[1]	-	-
Sodium monofluoroacetate	-	11 hrs in possums[2]	<11 h	7 days
Cyanide	-	6.5 min in possums[3]	+	12–24 h
Calciferol/ cholecalciferol	-	3-11 days in mice[4] 2-10 days in rat[5]	10-68 days	3 months
Zinc phosphide	2.5% in bait	4-5 hrs in bandicoot[6]	+	12–24 h
Brodifacoum	0.64 mg/kg for 4 days	4 days in rats[7]	130-350 days	24 months or longer
Bromadiolone	10 mg/kg	5-11 days in mice and rats[8]	-	-
Coumatetralyl	34 mg/kg for 4 days	6.5 days in rats[7]	50-70 days	4 months
Pindone	99.27mg/kg for 4 days	4 days in rats[7]	2.1 days	4 weeks
Warfarin	6.57mg/kg for 4 days	3 days in rats[7]	7-10 days[9]	24 weeks[10]

*After sub-lethal exposure; + = No published value but likely to be <12 h.

(Adopted and modified from Eason *et al.*, 2008 and Eason *et al.*, 2011)

[1]Marks *et al.*, 2004; [2]Littin *et al.*, 2010; [3]Gregory *et al.*, 1998; [4]Hatch and Laflamme, 1989; [5]Saini and Parshad, 1992; [6]Htun and Brooks, 1979; [7]Fisher *et al.*, 2004; [8]Meehan, 1978; [9]Thijssen, 1995; Fisher *et al.*, 2003.

3. Other less used toxins

The alkaloid, strychnine is the prime candidate for mountain beaver control (Campbell and Evans, 1988). Bjorge and Gunson (1985) reported that total mortality of cattle dropped significantly after wolf control using strychnine in Alberta. Bremner (1946) used strychnine-poisoned eggs as very successful for animal control. Strychnine (65 mg), stirred into a small hole in the side of an egg and one or two eggs placed in an active den about 2 ft from the entrance proved promising. Strychnine is now prohibited for control of most predatory animals, including badgers. Thallium compounds were introduced in 1920 for control of rodents, particularly prairie dogs and ground squirrels that refused to take strychnine baits. Arsenic tracking powders have also given excellent commensal rodent control in situations where its exposure did not constitute a hazard to other life (Crabtree, 1962).

Another toxin, sodium nitrite, that reportedly causes a quick and humane death, palatable yet toxic to wild pigs, cheap and accessible, is degradable in the environment and reduces risk to the operator as has an effective antidote was used in pig management programmes (Cowled *et al.*, 2008). Alpha-chloralose can also act as a very effective rodenticide for indoor uses (Mason and Littin, 2003).

Reducing risks to native and non-targets is also important for a successful chemical control of vertebrate pests. Use of bait stations restricts direct access of non-targets to baits, since the stations are modified to assure ready access to target species while minimizing access by non-target animals, especially birds. Ground based poisoning operations along forestry roads or tracks are cost-effective and cause minimal risks to native wildlife compared to aerial application (Alterio, 2000). Factors such as bait colour, delivery method, use of flavourants and timing of operation should be optimized to allow maximum impact on target populations while minimizing casualities among non-targets (Jackson and vanAarde, 2003). Knowledge on animal group locations as assigned by 'Judas technique' is used to refine the placement of toxicant baits and thereby increasing effectiveness and decreasing cost (McIlroy and Gifford, 1997). They identified the exact places of keeping warfarin baits for feral pigs, *S. scrofa* using 'Judas pigs' which resulted in control of even the low-density populations present after a larger control exercise two years earlier. Sub lethal doses of most of the toxicants used in animal management led to bait shyness. So sustained control of animal pests are unlikely to be achieved by repeated use of any of the toxins, but changing to other types with different modes of action will provide more effective long-term control.

8.5. Poisonous Gases and Fumigants

Fumigation is a method of pest control that completely fills an area with gaseous pesticides or fumigants to suffocate or poison the pests within. Poisoning with gas or fumigation usually involves the following steps, firstly the area to be fumigated is usually covered to create a sealed environment; next the fumigant is released into the space to be fumigated; then, the space is held for a set period while the fumigant gas percolates through the space and acts on and kills the pest, next the space is ventilated so that the poisonous gases are allowed to escape from the space and render it safe for humans. Different poisonous gases are used in pest management programmes with some success.

(1) Carbon bisulphide

Carbon bisulphide is used for the control of squirrels and rodents. The type of treatment used included the 'waste ball' method, which is still being used in some areas. This method consists of saturating jute balls (~ 2 inches dia.) in a bucket with carbon bisulphide. One ball is thrown deep into the burrow and the entrance is closed with a shovelful of earth and quickly trampled to insure air tight sealing. Another method is by exploding the gas in the burrow. The common procedure is to place a saturated waste ball in the burrow, seal the entrance with a piece of sod or a loose clod of earth for five to fifteen minutes to allow the gas to vapourize, then remove the clod and apply a small lighted torch to initiate explosion. The explosion will be extremely vigorous and has the advantage of indicating whether all the entrances of the burrows have been previously closed. This method should never be employed when the grass is dry as serious fires could result. Also extreme care should be taken when lighting the gas in the burrow. This procedure of firing the burrows is often used in orchards with no detrimental effects on the trees (Dana, 1962). Commercial pumps for the application of carbon bisulphide in furrows and holes are available. Bremner (1946) found that carbon bisulphide, although better, gave only about 50% control and the material had to be applied in spring while the soil is moist enough to retain the fumigant and not much risk of fire. Thus carbon bisulphide application is limited to the wet season; it is slow to convert into a gaseous state, slow to diffuse throughout the burrow and has a slow physiological effect on the animal. The Demon Rodent Gun® has removed some of these objections as it vapourizes the carbon bisulphide in a gaseous state, thereby quickly raising the concentration to a lethal level and preventing the squirrel from escaping so easily.

(2) Phosphine

The use of aluminium or magnesium phosphide tablets to create phosphine gas that asphyxiates pest animals in holes is a valuable method of pest management. Burrowing rodents on airports, such as woodchucks (ground hogs) and prairie dogs, can be killed by fumigating the burrows with aluminum phosphide tablets. Aluminum phosphide pellets react with moisture in the burrow to produce phosphine gas and are used for rodent control also (Cleary and Dolbeer, 2005). Aluminum phosphide pellets, when tested against mountain beavers, *A. rufa* resulted in 100% mortality and very effective (Campbell and Evans, 1988). Feral cats, *F. catus* are highly sensitive to phosphine gas, having a 30 min lethal gas concentration of 80 ppm, compared to 2400 ppm for rabbits (Campbell *et al.*, 2011).

(3) Methyl Bromide

The fumigant, methyl bromide has proved to be effective in the control of rodents as well as the insect pests. There is no danger from fire and it can be used in wet or dry soils and at various temperatures. Berry (1938) reported successful results in killing ground squirrels by injecting methyl bromide in burrows. However, methyl bromide has been identified as a significant ozone-depleting substance, resulting in regulatory actions (Thomas, 1996).

(4) Other Fumigants

A common method of controlling small infestations of mammal pest is the use of the exhaust of an automobile. One end of a hose may be attached to the exhaust pipe and the other end inserted in the burrow. Some of the many fumigants which have been tried are hydrocyanic acid gas, calcium cyanide, fumes of sulphur, gasoline, petroleum distillate, chloropicrin, kerosene and tetrachloroethane with varying success against rodents (Dana, 1962). Smoking and gassing of burrows, dogging with cyano gas or chloropicrin is used as a method of rabbit control in New Zealand (Howard, 1964). Tear gas, chloropicrin, has also shown some promise in situations of rat and mouse problems. At high concentrations the gas is lethal, but at lower levels it reduces activity or causes abandonment of the area (Tigner and Bowles, 1964). But chloropicrin slowly released in an orchard over a period of time from pressurized containers failed to prevent deer from rubbing their antlers on orchard trees (Welch, 1967). Chloropicrin was earlier used as a warren fumigant for European rabbits, *O. cuniculus* also

(Marks, 2009). Vapours of acrolein are quite toxic to mammals and are used as furrow fumigant (O'Connell and Clark, 1992).

Pyrotechnic Fumigants

A pyrotechnic fumigant is a device containing ingredients that emit toxic gases, smokes and fumes when burned. The pyrotechnic fumigant known as gas cartridge is used to control burrowing rodents specifically woodchucks, prairie dogs, gophers and ground squirrels. These gas cartridges contain sulfur, charcoal, red phosphorus, black summer oil, sodium nitrate, sawdust, borax and Fuller's earth as its ingredient. A two ingredient cartridge containing 65 g of 65% sodium nitrate and 35% charcoal was effective in both laboratory and field tests on wild Norway rats, *R. norvegicus*. In field tests conducted at a rat-infested cattle feedlot, there was a 77% (range: 35-95%) reduction in numbers of reopened burrows after fumigation as compared to pretreatment figures.

When carbon and sodium nitrate are burned carbon monoxide (CO) is generated according to the following formula.

$$4C + 2\,NaNO_3 \rightarrow 3\,CO + Na_2CO_3 + N_2$$

carbon + sodium nitrate → carbon monoxide + sodium carbonate + nitrogen.

Sodium nitrate and charcoal are not dangerous chemicals. The acute oral LD_{50} of sodium carbonate (soda ash) is 4,000 mg/kg (Frank, 1948) and charcoal and sodium nitrate is >3,000 mg/kg in rats and there was no potentiation when given in combination at 3,000 mg/kg. The lack of secondary toxicity to bobcats, *Lynx rufus* fed with rats killed by fumes from burning sodium nitrate and charcoal implies that carcasses killed by fumes from the cartridge would not be toxic to scavengers.

Gas cartridges or 'smoke bombs', as they are sometimes called have been used for the control of marmots, ground squirrels, prairie dogs, foxes, badgers and coyotes (Anderson, 1969). When ignited, the gas cartridges produce poisonous gases, including carbon monoxide and oxides of sulfur, nitrogen and phosphorus. Death may also be due in part to suffocation. Ammonium nitrate-diesel explosives poured into holes drilled into nest cavities of mountain beavers were used to blow up nests (Campbell and Evans, 1988). Methyl bromide and carbon bisulfide are effective fumigants for badgers and have been used extensively for ground squirrel control. Cyanide dust forced into badger holes which were then tightly packed with soil failed to control them because the badgers block the tunnel to avoid the toxic gas. Carbon dioxide (CO_2) is used in some enclosed indoor sites, especially cold stores (Meehan, 1984). At concentrations above 40%, it kills by causing a lack of oxygen (anoxia), leading to the loss of normal brain function and eventually respiratory failure in animals (Mason and Littin, 2003).

Many fumigants have been tested to find a gas that would be fatal more quickly and at low concentrations. It is also necessary for the gas to be safe to handle with reasonable precautions. Further the gas introduced should fill the rodent burrow in required concentration in short time to make pest control successful. For this reason, materials such as solid calcium cyanide which generate gas rather slowly are not usually effective. Complex runways and pockets in burrows, the hibernation of animals are the causes of some of the difficulties attending the use of gas. Another factor that makes the use of gas difficult is the habit of plugging against an enemy. So gases of low toxicity that diffuse slowly through the burrow, such as carbon dioxide and sulphur dioxide, have not proven successful. Though some fumigants and gases are found effective care should be taken well in advance to avoid hazards during treatment.

Chapter 9

Biological Management

Any living organism especially of these animal and bird pests are regulated in the environment by bottom-up (food supply) and top-down (predators, parasites) strategy apart from weather and environmental circumstances. Small animals are generally controlled by top-down strategy rather than bottom-up and *vice-versa*. Biological control is the use of living organisms as pest control agents (Waage and Greathead, 1988) which include parasites, predators and competitors. Management of vertebrate pests using predators and disease causing agents though effective in some conditions should be used only after thorough study on the ecosystem and with utmost care because the bio-control agent once introduced in the field is difficult to take back if the consequences are adverse. Besides this, the use of guard dogs to deter crop damaging deer and honey bees to deter crop raiding elephants are reported to be effective.

9.1. Guard Animals

Guard dogs

The concept of using dogs to guard farm and livestock from predators is an ancient one and can be traced back to many centuries B.C. in Eurasia (Bordeaux, 1974). Guard dogs were first brought to the Americas with the Spanish conquistadors to protect their flocks (Lyman, 1844). These guard dogs were commonly seen with large flocks of sheep in South America (Darwin, 1839). Apparently, the Spanish dogs did not persist into the 20th century as a result of inbreeding with indigenous Indian dogs (Black, 1981). But guard dogs are still being used and there are a dozen or so traditional guard dog breeds, they share the same basic behaviour. These dogs are bred to be large, placid, powerful, courageous and loyal. Breeds like Great Pyrenees, Komondor, Akbash, Maremma (Linhart *et al.*, 1979), Anatolian Shepherds, Shar Planinetz, Anatolian/Shars are used for protecting livestock (Coppinger *et al.*, 1988). Border collies are also used, as they are working dogs bred to herd animals' thus avoiding predator attack and respond well to whistle and verbal commands.

A single Border collie and its handler can keep an area of approximately 50 km² free of larger birds and other nuisance wildlife. Although they are effective at deterring ground foraging birds such as waders and wildfowl, they are not so useful for species that spend most of their time flying or perching, such as swallows (Erwin, 1999). Trained dogs can be used to flush birds to expose them to raptors or other hazing methods but that might not be highly effective if birds refuse to fly (Cooper, 1970; Lefebvre and Mott, 1987). The dogs especially Border collies were found efficient in deterring geese on the lands until they flew away or entered the water. The dogs again go into the water and further stalked geese until they flew away (Preusser *et al.*, 2008). Border collies were reported to drive away Canada goose from a property of 44 ha and reported as a very effective and economical management for the pest birds (Castelli and Sleggs, 2000). Guard dogs, *Canis lupus familiaris* were found effective in harassing and wading off wild turkeys, *M. gallopavo* from vineyards (Coates *et al.*, 2010). Mere presence of the dog was often sufficient to deter depredation of birds. They can also be released from periodic patrol vehicles to frighten birds.

Dogs either tethered or running free can be utilized to frighten deer, small paddocks etc (Howard, 1967). They also can be tethered on long running lines (Marsh *et al.*, 1991). Farmers described barking, marking with scent posts and actually chasing rabbits, monkeys, porcupines, deer, antelopes, coyotes, fox and birds from their farms. Some dogs would not tolerate eagles or other birds, cattle or even strangers. But the nature and extent of depredation reduction attributed to the dog's abilities was often difficult to describe. The exact method of depredation prevention by dogs is by barking and running towards the predators but only time will indicate if these animals and birds will learn to circumvent the guard dogs (Pfeifer and Goos, 1982). Dogs are used effectively by farmers of India to prevent crop damage by chasing the langurs, *Semnopithecus entellus* away from farm lands. Many times these dogs kill the langurs, in particular juveniles and infants (Chhangani and Mohnot, 2004). Killer dogs are used to control wild goats and rabbits when seen in dense bushes (Howard, 1964; Johnson, 1964).

In a study, restrained dogs failed to keep deer out of agricultural crops because deer became accustomed to the dogs and their barking (DeGarmo and Gill, 1958). However, free ranging dogs, confined by a buried fence and electronic collars were effective in reducing deer damage to a white pine, *Pinus strobus* plantation (Beringer *et al.*, 1994). Dogs of pure breeds (Border collie, Siberian husky) or of blood line mix (Labrador retriever mix, Hound mix, German shepherd mix, Malamute mix) were used to protect crop from wild life damage. Among these, Border collie, Siberian husky and Malamute mix were found effective to deter deer depredation and patrolling. It was reported that, after the recruitment two Siberian huskies, no damage occurred due to deer in vegetable plantations in an area of 1.4 ha. (VerCauterern *et al.*, 2005a). Paired dogs of Labrador and one Labrador-cross in one experiment, two husky/collie crosses in the second and two German shepherd in the third were used to deter deer from apple orchards. The mean yield per mm trunk diameter was higher in dog-protected than control plots by 21% in the first season and by 115% in the second. Similarly, mean yield per ha for dog-protected areas was higher than control plots by 37% and 128% with high economic returns of 348% and 404% compared to control in first and second seasons, respectively. Dog-protection had a profound effect on the growth of young trees also. Protected trees were 21% taller than controls in the first year and 61% taller in the following year (Curtis and Rieckenberg, 2005). Effectiveness can vary among breeds (Green and Woodruff, 1990) and likely is dependent on other factors including: 1) how the dogs were raised, 2) the habitat and topography of the area 3) the density and type of predators, 4) the availability and type of prey, 5) the number of dogs used, 6) other methods used to manage predation. The interaction and potential for synergism among these and other factors make it difficult to accurately predict the effectiveness of a dog. Guard dogs should be of larger size with a heavy coat so they could withstand winter conditions (Beringer *et al.*, 1994) and exhibit a propensity to bark and chase (Curtis and Reickenberg, 2005).

Studies investigating the efficacy of guard dogs have shown the dogs to be effective in some situation and ineffective in others (Linhart *et al.*, 1979). Some risk and problems when using guard dogs: (1) strangers to the farm lands had to be warned of the dogs, (2) some time dogs wander into neighbours' property, (3) some dogs injure pet animals (4) some times dangerous to owners also (5) incidence of accidental kill or theft of dogs from farms, (6) traps and snares used in the farm has to be modified to avoid capturing the dogs and (7) there is no assurance that the dog would have the instinctive ability to guard the farm even when properly raised. The dogs must respond well to voice commands or to whistles. The larger and faster breeds of dogs seem preferred and several dogs may be needed so as not to overwork them. Dogs would likely be of limited value for hazing birds from agricultural fields because they could only be used in orchards or low-growing crops and where the dogs could easily access the field and where the dogs themselves did not damage the crop. In some circumstances where fields are small, they may be useful if restricted only to the periphery of the field (Marsh *et al.*, 1991). Birds may also rapidly habituate to their presence, often moving a short distance away when the dog approaches but not leaving the area needing protection (Mattingly, 1976). Purchase price as well as the costs of transportation, maintenance and replacement constitutes a substantial financial investment (Green *et al.*, 1984). Replacement cost is also an important factor when considering longevity in working dogs. Lorenz *et al.* (1986) reported 50% of dogs working on farm/ranches died by 38 months of age and 50% of dogs on ranches died in their first 18 months. Though there are many risks, guard dogs are being used as a nonlethal, environmentally acceptable and economical method of reducing depredation. Apart from dogs the use of llamas, donkeys, maremmas and alpacas to guard livestock is also reported (Jenkins, 2003).

Llamas

Use of llamas for protecting livestock from predators takes advantage of their evolution with predators and defensive capabilities. Using llamas as guard animals is growing in popularity and found most practical and effective tool to deter predators mainly coyotes and wild dogs (Meadows and Knowlton, 2000). Traits that may be useful in selecting llamas as guard animals is they do not need a special feeding program, easy to handle, live longer than guard dogs (Knowlton *et al.*, 1999), their leadership trait, alertness and body weight (Cavalcanti and Knowlton, 1998).

Donkeys

Guard donkeys are used to protect livestock (Green, 1989) because they will bray, bare its teeth, chase and try to kick and bite coyotes and dogs (Acorn and Dorrance, 1998). Recommendations on the selection of donkeys as guardians include using only a jenny or gelded jack. Donkeys should be introduced to the livestock about 4-6 weeks prior to the onset of anticipated predation events to properly bond with the group and they are found most effective in small and fenced pastures (Acorn and Dorrance, 1998). Effectiveness of donkeys as guard animals is reported highly variable and usually not as successful as guard dogs (Green, 1989; Walton and Field, 1989).

Falcons

Trained falcons and other birds of prey have been used intermittently on various airports in Europe and North America to disperse birds since the late 1940s (Erickson *et al.*, 1990). Peregrine falcons (*Falco peregrinus*), gyrfalcons (*Falco rusticolus*) and goshawks (*Accipiter gentilis*) are the species most frequently used (Blokpoel, 1976). Heighway (1970) reported that a team of eight peregrines flown by two full-time trainers took two years to establish control over the resident gull populations. A team of four goshawks was successfully used at an airbase in Holland to clear the runways of gulls and the pest bird showed no signs of habituating to the goshawks during the two year study (Mikx, 1970). On the other hand, Hahn (1996) reported the use of falcons in German air fields as not much successful. The efficacy of trained hybrid falcons (*Falco* spp.) or hawks (*Buteo* spp. and *Parabuteo* spp.) at deterring scavenging gulls and corvids from landfills and found falcons as more effective then hawks (Baxter and Allan, 2006). In a study, goshawks were unsuccessful in deterring woodpigeons from Brassica fields (Inglis, 1980). After repeated attacks by the goshawk, the pigeons usually resettled and continued to feed. The advantage of falconry is that the birds on the airport are exposed to a natural predator for which they have an innate fear. The disadvantage is that a falconry program is often expensive, requiring a number of birds that must be maintained and cared for by a crew of trained, highly motivated personnel (Cleary and Dolbeer, 2005). Finally, falcons cannot be flown during bad weather such as fog, heavy rain or high winds.

9.2. Honey Bees

Bees are kept for the honey and wax that they produce and for the crops that they pollinate. Human use of honey bees in fighting also has a long history. Some of the earlier battles were fought with honey bees being the chief agents of victory. In the eleventh century, Irnmo, General of Emperor Henry I of England, threw bee-hives from cliffs onto the attacking troops of Duke of Lorraine (Karidozo and Osborn, 2005). Citizens of Gussing, Hungary used the same technique in 1289 against troops of Duke of Austria (Bromenshenk, 2004).

The local knowledge that the elephants run away from bee swarms (Vollrath and Douglas-Hamilton, 2002) had triggered off much discussion on whether or not bees could be used as 'guardians' to protect crops and trees from the destruction caused by elephant foraging. This was tested out scientifically by Vollrath and Douglas-Hamilton (2002) and revealed that under experimental conditions, acacia trees with beehives (either occupied or empty) did indeed have the effect of preventing elephant foraging damage to the trees. Karidozo and Osborn (2005) tested the effectiveness of bees in deterring elephants by comparing the mean severity of elephant crop damage in the test plots of maize, sorghum and cotton with honey bee hives. The results indicate that the presence of bee hives with or without honey bees did not provide the treatment plots with significant protection. Elephants avoided entrance points into the plots with live bee hives and they change the paths where bee hives are kept and entered through areas where bee hives are not kept.

In Kenya, they deployed a 90 m fence-line of nine inter-connected hives, all empty, on two exposed sides of a square 2 acre farm that was experiencing high levels of elephant crop depredation. Compared with a nearby control farm of similar status and size, the experimental farm experienced fewer raids and consequently had higher productivity thus suggesting a 'guardian beehive-fence' (King *et al.*, 2009). This bee fence and its efficiency was demonstrated later in Kenya with 1700 m of beehive fences semi-surrounded the outer boundaries of seventeen farms and compared elephant's farm invasion events with these and to seventeen neighbouring farms whose boundaries were protected only by thorn bush barriers. Out of 45 successful farm invasions, 31 invasions occurred through thorn barriers and only one through a beehive fence. A further thirteen attempts to enter a farm were deterred by the beehive fences and in five of these cases, the elephants walked along the entire line of the beehive fence before breaking through a thorn bush barrier. These results demonstrate that beehive fences are more effective than thorn bush barriers at deterring elephants. Additionally, the harvesting of 106 kg of honey during the trial period suggests that beehive fences may also improve crop production and enhance rural livelihood through honey sales (King *et al.*, 2011).

Using honeybees to deter elephants from raiding crops is fraught with many challenges. It is important to note that either a technical or a practical perspective, being able to use bees on a large scale is questionable especially when it happens to protect lengthy crop boundaries. Crop raiding is nocturnal and at that time bee activity will be at its lowest. Moreover they will not fly during wind or rain and at low temperatures. Considering all these, only the sound of disturbed bee may find a choice in deterring elephants. The sound of disturbed African honeybees, *Apis mellifera scutellata* causes African elephants, *L. africana* to retreat and produce warning vocalizations that lead other elephants to join the flight (King *et al.*, 2007).

9.3. Human Patrols

Patrols by humans on foot or in vehicles have long been used for hazing or frightening birds and animals from agricultural fields. The presence of humans is one of the most effective animal deterrent available, as well as being one of the simplest. It is generally accepted that it reduces damage, but it is unclear whether it is cost-effective as a sole control method. But this forms the most common method of animal and bird control used in developing countries. Women and youths, 'bird boys' are either employed or members of the farming family take it in turns to scare birds and animals from their farms (Akande, 1978). The basic and simple way of frightening birds involves standing in full view of the birds and raising and lowering the arms in mock wing beats of about 25 per min (Wright, 1969).

Humans on foot, cycles, horseback or in vehicles may intentionally or inadvertently cause birds to flee (Owens, 1977; Kenward, 1978; Burger 1981). Kenward (1978) examined the influence of human activity on woodpigeons, *C. palumbus* feeding in brassica fields and found effective. Vickery and Summers (1992) reported that brent geese, *Branta bernicla*, rarely attempted to visit sites where intensive human disturbance had been applied. There were no signs of habituation to the human scarer and damage to the crop was reduced by an order of magnitude. Although highly effective, this is an expensive and labour intensive deterrent and a judgement must be made regarding the losses incurred through crop damage compared to the cost of using a human scarer. Human patrols are generally used in combination with other techniques, such as shooting or firing cracker shells, to provide variety in an integrated hazing program. The presence of humans is probably best utilized to reinforce the danger associated with other frightening techniques during their operation or servicing. Patrols by humans usually include the operation of bird frightening devices such as shooting live ammunition or shell crackers or broadcasting bird distress/alarm calls. Depending on the bird species and situation, birds may be already accustomed to people or rapidly habituate to human presence unless it is occasionally reinforced by shooting or other means. Trained dogs can aid in the effectiveness of human patrols. Trained dogs, for example, can be used to flush birds to expose them to raptors when deployed from vehicles.

Bird reactions to people varied with the human activity. People walking along the beach or jogging on paths always disturbed birds when they were present. Birds often ignored horseback riders and people working in farm without much noise even when they came within 18 to 20 ft. In general, small birds and those in suburban areas allowed the closest approaches before flying. Cooke (1980) speculated that suburban birds were more accustomed to human presence than those in rural areas. Reactions vary among species, however, may rapidly habituate or if approached too closely, move only a short distance away and return soon after the people depart. Human patrols are not highly effective if birds refuse to fly (Cooper, 1970; Lefebvre and Mott, 1987).

Systematic patrolling in the early morning and late evening effectively reduce jack rabbit damage in localized areas (Johnson, 1964). Wild boar damage is reported negatively correlated with the grade of anthropogenic activity (Schley *et al.*, 2008). Cacao crop with human guards making noise during the harvest season had fewer losses due to squirrel (*Sciuridae* sp.), chimpanzee (*Pan troglodytes*), agile mangabey (*Cercocebus agilis*) and moustached guenon (*Cercopithecus cephus*) (Arlet and Molleman, 2010). The most commonly used crop protection strategy in guarding agricultural fields from damage by langurs, *S. entellus* is by constant vigilance during crop reasons. Many of the guards use 'gophan', a device to throw stones catapult towards invading langurs to chase them away from the field. For successful guarding it is required that people be in the fields during the seasons when the crops were most vulnerable and this guarding should have to be throughout the day. Obviously, this was not possible because people had many other works to do. Other studies have also reported that primates are more fearful of adult man than of women and children (King and Lee, 1987; Hill, 2000).

The most common protection strategy for farmers is to guard their fields against nilgai and blackbuck by remaining vigilant during the crop season (Chauhan and Singh, 1990). Guards stationed overnight in two watchtowers and using powerful torches to scan the area for approaching elephants has decreased elephant raids by 89.6%, compared with the group of control farms (Sitati *et al.*, 2005). Human activity can have a substantive impact on wild pig behaviour, movement and survival (West *et al.*, 2009). Even for livestock protection, human presence as a good herder who is able to stay with and monitor livestock can be an effective method of protection (Linnell *et al.*, 1996). Furthermore, humans are able to observe when predators enter an area, employ aversive or disruptive stimuli and identify the characteristics and timing of predators and predation. It is possible to maintain a human guard who walks through the field watching for and chasing away wildlife (Shivik, 2004).

9.4. Shooting and Hunting With Dogs

Under certain conditions, shooting has proven effective means for reducing vertebrate pests of crop plants. A great deal of shooting is done, but considered to have only a small effect on pest bird numbers. As a scaring method, however, shooting can be highly effective (Seubert, 1964). At times shooting is regarded as the only way to deter animal pests especially of large animals. Large mammals can be captured with tranquilizer guns, but this is generally not a practical or desirable or economical option (Cleary and Dolbeer, 2005). Sport hunting typically results in the removal of mostly adults and this alone may not be enough to reduce the population (Bieber and Ruf, 2005). Shooting is especially useful in controlling animals with low reproductive rates, such as porcupines.

Shooting is a selective method which is reported to be effective in the control of the Lewis woodpeckers, *Melanerpes lewis*. Both the shotgun and the .22 rifle have been tried and it was reported that the shotgun is more effective for bird control (Koehler, 1962). Murton *et al.* (1972) tested the effect of shooting woodpigeons, *C. palumbus* damaging cabbages and sprouts. In each of two years a shooting site and a no shooting site were compared but resulted in insignificant difference in damage between the sites in either year rendering it ineffective. Another study of control of woodpigeons reported that shooting of the birds around roosts killed large numbers of birds but it did not increase winter mortality above that in the absence of shooting (Murton *et al.*, 1974). In a study, shooters killed more than 26,000 laughing gulls and 2,300 other gulls flying over the airport in 2,206 person-hours of shooting but shooting did not appear to condition gulls to avoid flying over the airport but was found to substantially reduce the incidences of strikes between all species of gulls and aircraft, by 70% in 1991 and 89% in 1992 (Dolbeer *et al.*, 1993). A program to reduce gull strikes was conducted from 1991-2002 in which 2-5 people stationed on airport boundaries shot gulls flying over the airport. As a result, the number of strikes with laughing gulls was reduced to 38% during 1991 and 24% during 1992-2002 when compared to 1988-1990. Strikes by the 3 other gull species were reduced to 24-52%. The laughing gull colony in the nearby Jamaica Bay has declined 58% from 7,629 nests in 1990 to 3,238 nests in 2002. The colony size declined by only 58% from 1990-2002 while the annual strike rate of laughing gulls declined by 97% indicated that many laughing gulls altered flight patterns in response to shooting to avoid the airport (Dolbeer *et al.*, 2003).

Organized rabbit hunts have proven satisfactory in reducing the population of the pest animal over local areas. This type of control is usually more effective in open terrain rather than cultivated fields, orchards and vineyards; there has been reluctance by farmers and sportsmen's clubs to organize these hunts due to the responsibility or liability of injury or damage involved (Johnson, 1964). Night shooting was practiced in New Zealand to reduce rabbit populations (Howard, 1964). Some species like deer, feral pigs and Himalayan thar, *Hemitragus jemlahicus* are controlled normally by recreational (Nugent and Fraser, 1993) and commercial hunters (Parkes *et al.*, 1996). Hunters do hold thar numbers at the low densities but only in places where they have easy access (Forsyth, 1999). Hunting can effectively manage deer populations in rural (VerCauteren and Hygnstrom, 1998; Woolf and Roseberry, 1998) and urban (Hansen and Beringer, 1997; VerCauteren and Hygnstrom, 2002) areas and in some cases can be acceptable as the primary tool of deer population management (Brown *et al.*, 2000). Single shot guns and potash bombs are used against crop raiding langurs which are usually killed or seriously injured (Chhangani and Mohnot, 2004). Eradication by shooting of feral goats, *Capra hircus* in Raoul Island was described by Parkes (1990). A comparison of the efficiency of shooting buffalo, *Bubalus bubalis* in northern Australia either from the ground or from a helicopter was reported by Boulton and Freeland (1991). Choquenot (1990) reported shooting as a population management of feral donkeys (*Equus asinus*) in northern Australia. Later, Freeland and Choquenot (1990) and Choquenot (1991) reported that a nearby donkey population not subject to shooting had not increased significantly over the same time period. Persistent crop raiding elephant bulls can only be deterred by gunfire including shooting one in the group (Osborn and Rasmussen, 1995). It is generally believed that shooting an elephant at night while it is raiding was the best way to 'teach' the other elephants to stay away.

Hunting occurs both as a sport as well as a control measure of pigs in New Zealand (MacKintosh, 1950; Shennan, 1960). As pig populations decline, however, aerial shooting produces diminishing returns and probably is not cost-effective at low population densities (Choquenot *et al.*, 1999). In other areas, weather, heavy cover and rough terrain also limit the applicability of aerial shooting. Two types of equipments are available for night shooting *ie*, systems that use near-infrared light and systems that use thermal imaging. Since thermal imaging works solely by capturing infrared energy from an object, no light at all is required for the device to function. Warm objects stand out against cooler backgrounds and become visible even at a distance of 0.5 mile away (West *et al.*, 2009). Shooting programmes are found very effective once when the animal groups were located using 'Judas technique' and thereby increasing effectiveness and decreasing cost (McIlroy and Gifford, 1997). In poor habitats simple recreational hunting alone has resulted in low pig populations (Belden and Frankenberger, 1990) but usually ineffective as a population control method in good habitat (Hanson *et al.*, 2009). Hunting of foxes either for their pelts, a bounty or merely as a sport has long been seen by the agricultural community as a useful and economic way of regulating fox numbers. Den hunting of coyotes is very effective to stop predation on sheep (Till and Knowlton, 1983)

Hunting with Dogs

Dogs have often been used as a hunting and detection tool for rodents (Campbell *et al.*, 2011). Hunting with dogs can be an effective management practice to reduce wild pig populations in local areas and has been successfully used as part of larger control programs (Choquenot *et al.*, 1996). Dogs used for hunting are of two types *viz.*, dogs hunt by sight such as greyhounds and trail hounds. The former type is usually kept in box or cage and released when the mammal pest is seen to catch and kill them. The other usually follows the mammal by its scent. Several breeds of dogs like Bluetick, Walker and Redbone are used commonly for hunting with 2-5 dogs together. Trained trail hounds are used to catch raccoons, opossums, bobcats, bears and mountain lions. Another technique of fox hunting found in some parts of Australia is the use of small terrier dogs to flush foxes from dens. Dislodged animals are either killed with shotguns or coursed with large lurcher dogs. McIlroy and Saillard (1989) described an evaluation of hunting along with dogs, of feral pigs in mountain forests of south-eastern Australia. The hunters killed only 13% of pigs known to be in the area. In some cases, hunting with dogs simply causes pigs to move into adjacent areas. This shift in location can protect small, isolated, sensitive areas but may simply relocate the problem rather than alleviate it (Barrett and Birmingham, 1994). On the other hand, some have speculated that harassment, such as that created by hunting with dogs, can cause home range shifts away from particular areas of concern and thus is a viable management technique (Gaston *et al.*, 2008; Hayes *et al*, 2009). Though the dogs were found to enhance the success of hunters, increased use of hunting dogs is unlikely to have substantial effects in managing overabundant deer populations (Godwin *et al.*, 2013). Disadvantages of hunting are, it is time consuming, reductions are short-term (Purdy *et al.*, 1987) and assessment of efficacy is difficult (Erickson and Giessman, 1989). Hunting proved to reduce wild boar damage (Geisser and Reyer, 2004). However, wild boar reproductive rates can increase up to 200% under ideal conditions and therefore, populations can recover within one year period (Schlageter and Haag-Wackernagel, 2012).

9.5. Predators

Large top predators such as wolves and big cats have a significant effect in structuring terrestrial communities and maintaining species diversity in lower trophic levels. With removal of large predators, the herbivores may become over abundant and displace some species while destroying habitat for many others (Berger *et al.*, 2001; Terborgh *et al.*, 2001). For example, eradication of wolves, *C. lupus* and grizzly bears, *Ursus arctos* in Grand Teton National Park resulted in a fivefold increase in moose (*Alces alces*) density, with consequent over browsing of riparian vegetation and disappearance of migratory birds dependent on such vegetation (Berger *et al.*, 2001). The rates of increase of some marsupials indicate density-dependent predation by red fox, *V. vulpes* especially when good forest habitat provides refuge for the prey (Sinclair *et al.*,1998). Fox control attempted by baiting with sodium monofluoroacetate (compound 1080) in two experimental areas caused increase in population of rock-wallabies, *Petrogale lateralis* by 138% and 223% (Kinnear *et al.*, 1988). Thus foxes are said to have an important effect on rock-wallaby population dynamics.

Another mechanism by which top predators might contribute to maintenance of species diversity is through the control of mid-sized predators or mesopredators. In the absence of top-down control, mesopredators may increase in abundance, leading to intensification of predation pressure on smaller prey species. Thus a process of 'mesopredator release' in the environment can result in management of small prey species (Courchamp *et al.*, 2000). Some studies have demonstrated declines in prey as a result of mesopredator release in small habitat patches without large predators (Palomares *et al.*, 1995; Crooks and Soule, 1999). However, in an undisturbed ecosystem, large predators, mesopredators and small mammal preys coexist and some sort of natural balance between them on reproduction and predation makes a proper regulation of every organism at certain levels.

Pest control by natural predation

Natural enemies of the rabbits include the large hawks and owls, eagles, coyotes, wildcats, foxes, weasels, dogs and cats. But these natural enemies are seldom ever numerous enough to control rabbits. Even though other methods must often be used, these natural enemies should not be indiscriminately slaughtered. Raptors are reported to regulate small mammals in many cases (Erlinge *et al.*, 1983, 1988). Predation is said to exert a powerful influence on vole dynamics; survival and reproduction are reportedly low when predators are common (Norrdahl and Korpimaki, 1995; Klemola *et al.*, 1997), with predators capable of extirpation. Vole populations in the mainland of northern Europe are strongly limited by predation from specialist and generalist avian and mammalian predators (Erlinge *et al.*, 1983; Hanski *et al.*, 1991; Korpimaki and Norrdahl, 1998; Korpimaki *et al.*, 2002). Avian predators such as Eurasian kestrels (*Falco tinnunculus*), short-eared owls (*Asio flammeus*) and eagle owls (*Bubo bubo*) also control the vole populations. Minks are implicated in the demise of the water vole, *Arvicola terrestris* in Great Britain (Woodroffe *et al.*, 1990, Barreto *et al.*, 1998) and are said to have a greater impact in fragmented landscapes (Rushton *et al.*, 2000; Aars *et al.*, 2001; Telfer *et al.*, 2001; MacDonald *et al.*, 2002). In Fennoscandia, vole cycles are driven by avian and mammalian predators (Korpimaki and Norrdahl, 1998; Korpimaki *et al.*, 2002) while in island areas there is a negative correlation between adders, *Vipera berus* and vole numbers (Lindell and Forsman, 1996). Feral American mink, *Mustela vison* in Finnish archipelago of the Baltic Sea was found to have a strong impact on the abundance and dynamics of field voles, *M. agrestis* and bank voles, *C. glareolus* (Banks *et al.*, 2004). Rodent pests in pine plantations reported to be controlled by avian and mammalian predators in Chile (Murua and Rodriguez, 1989; Munoz and Murua, 1990). However, barn owls could provide only a limited reduction of prey rodents in Israel (Kahila, 1991).

It might be possible to 'buffer' predator numbers by supporting them, or altering the environment, e.g. cats on farms by feeding or owls in oil palms by giving nesting sites (Wood, 1985). Artificial perches were reported significantly increased raptor numbers in agricultural areas (Hall *et al.*, 1981) but with no measurable reductions in rodent pest populations in agricultural crops (Askham, 1990), nor of pest birds in vineyards (Howard *et al.*, 1985). Native predators may exert some control over prey species in certain situations, but they are unpredictable, and cannot be relied on to prevent or control wildlife damage (Howard *et al.*, 1985). Regulation of prey populations by predators occurs under special circumstances of predator behaviour (their functional response) and population dynamics (the numerical response) (Pech *et al.*, 1995; Sinclair *et al.*, 1998; Sinclair, 2003).

Introduction of predators for pest control

Introduction of predators in island ecosystem or isolated ecosystem for specific pest species has resulted in great success. Foxes and raccoons were liberated in an island where the herring gulls were using as a breeding spot and found reduce the pest bird by 95%. The predators were trapped again and removed after the food supply had been exhausted (Howard, 1967). Introduced predators in a localized situation, *e.g.*, house cats, can depress and keep depressed below environmental capacity, a confined population of rats or house mice for as long as the diet of the cats is supplemented periodically with other food. House cats are also reported to reduce resident house sparrow population by 80%. Farmers devise systems using predation to control house sparrows by building catwalks around the barn at rafter level. Scrap lumber can be used to provide farm cat's access to locations where sparrows usually roost or build nest. Once the cats were able to patrol the barn, the sparrows quickly vacate the building (Fitzwater, 1994). The Australian brush-tailed opossums were reported to be controlled to a certain extend by predatory dogs in New Zealand. Killer dogs are used to reduce feral goats in places with dense bushes in New Zealand (Howard, 1964). In Australia, rabbits are reported to be regulated by European foxes, feral cats and dingoes. Further, dingoes are considered as major predators regulating population of kangaroos and probably feral goats and pigs (Newsome, 1990).

Failures of predator release

As stated above, introduction of a predator onto a very small isolated locality might result in complete extermination of a certain kind of vertebrate prey, but in most situations an introduction of alien predators into a new ecosystem is not only perilous but may prove to be catastrophic, because the predators of vertebrate pests are mostly not host-specific. New Zealand's first Rabbit Nuisance Act established in 1876 paved a way to introduce weasels, stoats and ferrets with a belief that they would control the rabbits but not found successful (Thompson, 1958). During the period 1910-1930 an area of 600 ha of the Frisian island of Terschelling was planted with young trees. Much damage was done to these plantations by water voles, *A. terrestris terrestris*. For the biological control of this pest 9 stoats, *Mustela erminea* were introduced in 1931 and they increased strongly and had to be controlled in turn. They exterminated the water voles within 5 years and reduced the population of the rabbit, *O. cuniculus* on the island to an extremely low level. Soon the stoats commenced feeding on sparrows, starlings, terns, shellducks, curlews, other waders, poultry, tame ducks and even turkeys. After 1939, however, a state of natural balance seems to have established itself, the stoats, though still common, are by no means a pest anymore (Howard, 1967). Thus, introduction of predators having a non-specific diet could result in severe side-effects in new localities, especially if there were no effective native predators (Wood, 1985). Failures of introduction of predators in pest control are mainly because of either of the two reasons: (1) that predators are usually versatile and rarely obligate, particularly out of native habitat; and (2) that predation by one species alone is only rarely a population regulator of vertebrate animals (McCabe, 1966). Further, uncontrolled predation has reported to result in species extinction also. Foxes have driven some highly vulnerable Australian marsupials to extinction (Burbidge and McKenzie, 1989; Kinnear *et al.*, 1998; Sinclair *et al.*, 1998). For example, small populations of black-footed rock-wallaby, *Petrogale lateralis* on rock outcrops in Western Australia declined to extinction once they dropped below some lower threshold number as a result of fox predation. It is reported that predation by foxes leads to the extinction of three species of bettongs, *Bettongia lesueur,*

B. penicillata and *B. gaimardi* and decline of rufous bettong, *Aepyprymnus fufescens* and long footed potoroo, *Potorus longipes* in south Wales (Pech and Hood, 1998). Predation by brushtail possums, *Trichosurus vulpecula* and ship rats, *R. rattus* is considered as the most immediate cause of kokako, *Callaeas cinerea wilsoni* decline (Innes *et al.*, 1999).

Some reports say vertebrate predators usually do more to increase population densities of field rodents than they do to depress them. It implies that, without predation, self-limitation stress factors come into play at lower density levels and that these forces operate as population controls more drastically than does predation (Howard, 1967). Some mammals, like some birds, can be extraordinarily effective hunters. But even among these skilled hunters, the victims seem most frequently to be the handicapped - the immature, the wanderer and the ill-adapted. The removal of such prey may conceivably result in more vigorous prey populations rather than less. Tragic examples include introduction of the mongoose, *Herpestes anropunctatus* into Hawaii to control rats, the fox into Australia to check rabbits and New Zealand's introduction of weasels, stoats and ferrets in the mistaken belief that they would control the rabbit. All of these introduced predators not only failed to accomplish their mission but have themselves become troublesome predators. Rodents and hares, all decreased in numbers at the very time when 22 species of their avian and ground predators were thought to be controlled. Pocket gopher numbers showed no correlation with the presence or absence of coyotes (Robinson and Harris, 1960). Mongoose, *H. anropunctatus* were introduced to control rats, *Rattus* spp. in Jamaica during 1872 (Laycock, 1966). Although initially successful in reducing rat populations, mongooses were not particular about their prey and nearly extirpated four species of ground-nesting birds. Nearly 20 years later, the mongoose, originally considered very beneficial, came to be regarded as the greatest pest (Hygnstrom *et al.*, 1994). Similar unsuccessful introduction of mongoose for rat control in Hawaii is reported by Hone (1994). Foxes, *Dusicyon griseus* were introduced in 1951 as biological control agents of European rabbits, *O. cuniculus* in Tierra del Fuego. Analysis of fox diet suggested that the foxes had not and could not control the rabbit population (Jaksic and Yanez, 1983). European sparrow hawks took a large percentage of house sparrows but could not conclude that the population was impaired. Natural predation may be considered a beneficial service for some prey species and for a few species it may even be important for survival.

The use of natural enemies (predators) to control pest populations of vertebrates is not a simple procedure because the role played by vertebrate predators as enemies of pest vertebrates (*e.g.*, rodents, rabbits and birds) is not a phenomenon easily interpreted, even empirically. In general predators regulate the herbivory pest population but only when the pest population is lowered by some other means triggered climatically, by disease or by human intervention. In such cases, carnivores can control mammalian pests for long periods. When predators are themselves pests to be controlled, integrated pest management may be required to avoid unwanted resurgences of other pests (Newsome, 1990). Introduced predators can have dramatic impacts on native prey populations which often lack the behavioural traits to avoid alien predators and their populations have little resilience to additional heavy predation (Dickman, 1996). No doubt, the combined predation pressure by native hawks, owls, snakes, and carnivores usually establishes a greater, not lesser, seasonal and annual density of species of vertebrate prey than would otherwise exist. The weak point concerning the effectiveness of native predators in controlling pestiferous mammals and birds is that the predators concerned are not host-specific; in general usually being opportunists, taking what is most readily available. Indeed, island ecosystems have suffered the highest extinction rates for mammals and introduced predators are thought to have been a major cause (Courchamp and Sugihara, 1999). Predators can regulate prey if such preys are the primary food source and there are few suitable alternative prey species (Sinclair, 2003). But predators are also self-limiting, usually reproducing much more slowly than their prey and often prey on other predators. Furthermore, even when predators actually do temporarily depress a vertebrate prey population, they do not necessarily reduce it to a density acceptable to man's needs. Even excessive predation can be compensated for by accelerated reproduction (Howard, 1967).

9.6. Disease Causing Agents

There undoubtedly have been many attempts reported or unreported to control wildlife populations with disease causing agents. It is known that as early as the 1880's, Pasteur recommended the introduction of a bacterial pathogen to reduce the rabbit population in Australia; in fact his project was not approved by Australian Government (Howard, 1967). Anyhow, biological control especially with pathogens of invertebrates has been very successful while that of vertebrates has largely been unsuccessful with few exceptions like myxomatosis in rabbits. Even for the most potential pathogen, utmost care and exhaustive studies including modeling should be done before its introduction for management. The biosafety measures and non target effects are to be studied because, these organisms once introduced is difficult to retrieve. The impact of introduction on the ecosystem as a whole is to be considered for successful biological control using disease causing agents.

I. Pathogens of Bird Pests

There is a tremendous volume of published data on the occurrence of potential disease causing organisms in wild birds but none was found to create an epizootic for a particular pest bird (Herman, 1964). Botulism caused extensive losses among shorebirds, pheasants, poultry birds and waterfowls. *Salmonella* infections have been reported from starling, rusty blackbirds and cowbirds in New Jersey. These bacteria are pathogens of the intestines and cause disease, often fatal in a wide variety of animal hosts also. Encephalitis is a virus reported to cause mortality in English sparrows and pheasants. Pox, another virus infection frequently recognized in birds, is manifested by the development of tumors. Though these all are potential pathogens, the main point here is that none of these are host-specific to any pest bird and they can be expected to be found in at least a variety of passeriform birds if not in most species of birds (Howard, 1967). Avian influenza virus/ flu virus that cause the bird infection can mutate to infect other animals and humans. Thus none of the pathogens or any parasites are presently used in bird pest management programmes, providing a wide opportunity for exploitation and establishment.

II. Pathogens of Animal Pests

Myxomastosis

Myxomatosis of rabbit forms the classical example for a disease agent being used intentionally to control a wild animal population (Fenner and Ratcliffe, 1965). Myxomatosis virus was introduced to Australia in the 1940s and 1950s to control rabbits. Early research in the 1940s in South Australia and in 1950 in Victoria and New South Wales gave discouraging results and it was thought that the virus may be of limited value for rabbit control (Fenner and Ratcliffe, 1965; Fenner, 1983; Fenner and Fantini, 1999). Drastically, in late 1950 and early 1951 the virus apparently had a devastating effect on rabbit populations (Fenner, 1983). The results were truly fantastic with a spectacular decline in counts of healthy rabbits (Myers *et al.*, 1954) before the virus had become attenuated and appearance of genetic resistance in the rabbit population. Repeated epizootics have resulted in only less mortality of rabbits (Fenner and Ratcliffe, 1965). In Britain also, Ross and Tittensor (1986) reported that the rate of spread and the proportions of rabbits infected with and dying from myxomatosis had declined since the disease was first recorded in British rabbits in 1953. This was suggested to be because of higher immunity of the rabbit population, though it was apparently also related to the evolution of strains of differing virulence (May and Anderson, 1983; Dwyer *et al.*, 1990). Myxomatosis, the disease against rabbit pest which was so successful in Australia, did not work in New Zealand, presumably because there were not enough mosquitoes (Howard, 1964). Blood-sucking arthropods were found capable of transmitting the infection and mosquitoes being the chief vector. The vector has been referred to as a 'flying pins' because, its mouth parts become directly contaminated with the virus rather than by the virus developing or multiplying within its body. An evaluation of myxomatosis for rabbit control on the sub Antarctic Macquarie Island reported variable results (Brothers *et al.*, 1982). Myxomatosis had greatly reduced the rabbit population in Tierra del Fuego (Jaksic and Yanez, 1983).

Rabbit haemorrhagic disease

Rabbit calicivirus disease also known as rabbit haemorrhagic disease was reported to cause mortality in domestic rabbits, *O. cuniculus* in China (Liu *et al.*, 1984). Over the next decade, the virus was detected in Europe and Central America both in wild and domestic rabbits. Following the arrival of the disease in Spain the population of rabbit has reduced to 20-80% of their former abundance, thus considered as a potential biocontrol agent. This virus has been introduced in many parts of the world as a biological control for rabbits. Initial outbreak of disease in wild rabbits in Spain and Australia were reported to cause a reduction of 65-95%, population (Pech and Hood, 1998). The decline of rabbit abundance due to haemorrhagic disease was significant in Spain and Portugal and up to some extends in France whereas the virus did not reduce the pest by and large in Britain and Northern Europe. In New Zealand, rabbit population has reduced to 50% immediately after six weeks of release of the virus but after that variable results are being reported. The number of rabbit counts declined by 72% where the virus was deliberately spread on baits and by 87% in areas of natural spread (Cooke and Fenner, 2002). Mosquitoes and rabbit fleas are capable of transferring the virus to healthy rabbits (Lenghaus *et al.*, 1994). Further, it was reported that the disease has reduced abundance of rabbits particularly in dry regions. Rabbits in cooler areas and high rainfall zones could able to maintain the population in-spite of this disease (Cooke and Fenner, 2002). Introduction of rabbit calicivirus disease has reduced the rabbit population significantly causing subsequent reduction of native raptors which have rabbits as their primary food (Newsome *et al.*, 1997). The weed species which are controlled by rabbits were also increased substantially.

Hog cholera

A very successful project to eliminate the wild pig population on a privately owned island off the coast of California by the introduction of hog cholera virus was made. However, it was not much effective against native swine in another Island (Howard, 1967). Nettles *et al.* (1989) reported use of swine fever (hog cholera) on islands off California for biological control of feral pigs. The abundance of feral pigs apparently dropped rapidly but the disease did not persist for a long time. That may have been because the abundance of feral pigs, after disease introduction was reduced to below the threshold host abundance for disease persistence or because all the susceptible pigs that survived became immune. The evaluation was limited because of the lack of accurate or precise data on abundance of feral pigs before or after disease introduction.

Feline parvo virus

Feline parvo virus causing feline panleucopaenia (FPL) in cats was introduced to Marion Island in the southern Indian Ocean in 1977 to control feral cats, *F. catus* (vanRensburg *et al.*, 1987). The disease is found host-specific, highly contagious and causes high mortality amongst kittens. The cat population on Marion Island dropped from an estimated 3409 in 1977 to 615 in 1982, suggesting an annual rate of decrease of 29%. The estimated reduction of feral cats was about 80% over 5 years of introduction of the disease (Howell, 1984).

Bacterial disease in rodents

Davis and Jensen (1952) reported on experimental attempts to introduce an epizootic among wild rats by inoculating a bacterium, *Salmonella enteritidis*, which is considered highly pathogenic to rats. They concluded that the organism is highly potential to cause pathogenicity but have no measurable effect on population size, mortality or reproduction. Only limited success has resulted from a commercial bacterial rodenticide, Ratin® but used extensively in Russia for the control mice and voles (Bykovskii and Kandybin, 1988).

Protozoan on rodents

The host specific protozoan, *Sarcocystis singaporensis* is found endemic on rodent populations in south Asia. It uses snakes, *Python reticulatus* and rodents of the genera *Rattus* and *Bandicota* to maintain its life cycle (Wood, 1985). Bait-pellets containing high numbers of *S. singaporensis* were given to rodents of three species (Norway rat, *R. norvegicus*, Malayan field rat, *R. tiomanicus* and greater bandicoot, *Bandicota indica*) in different agricultural habitats (chicken farm, oil palm plantation, rice field) caused a mortality ranging from 58% to 92% in Thailand. Detection of merozoites of *S. singaporensis* in lung tissue samples of rats collected dead at the experimental sites using a species-specific monoclonal antibody confirmed that protozoan as the causative agent of mortality thus formed as an effective population control measure (Jakel *et al.*, 1999).

Nematode on mice

Investigation on the potential of the nematode, *Capillaria hepatica*, as a biological agent in the control of wild mice in the cereal-growing regions of Australia showed a positive response. Singleton and Spratt (1986) studied the effects of *C. hepatica* on fecundity in laboratory mice and reported significantly lowers the litters produced and the number of young weaned per female infected mouse compared with uninfected. Another nematode, *Heligmosomoides polygyrus* was also found effective against mouse particularly the woodmice. An elegant series of experiments on the effects of *H. polygyrus* on laboratory colonies of mice provide strong evidence that this parasite can regulate mouse abundance (Scott, 1987; Scott 1990). Subsequent studies in outdoor enclosures examined the interaction between *H. polygyrus* and woodmice, *Apodemus sylvaticus* and found to have a significant impact on the survival of woodmice (Gregory, 1991).

Other pathogens

Many other disease causing agents are also being considered for the management of animal pests. Ectromelia virus which causes mousepox can also be considered as a suitable pathogen to control mouse populations (Buller *et al.*, 1986; Shellam, 1994; Jackson and vanAarde, 2003). This virus is reported to kill 80% of infected mice (Fenner, 1949) though some populations exhibit innate resistance to ectromelia (Buller *et al.*, 1986). A helminth species which reduces host survival or fecundity at an increasing rate as host abundance increases has a role in host population regulation (Spratt, 1990). The trypanosome of rabbit, *Trypanosoma nabiasi* can also be considered for the control of rabbits (Hamilton *et al.*, 2005).

Conclusion

There are many pathogens that cause disease in birds and animals but none found most useful in the point of pest management with a few exceptions. The largest impediment to progress in this line is a dearth of high quality research (Singleton, 1994). Coevolution is an important phenomenon in host-parasite or host-pathogen associations; nevertheless they may limit host population abundance (Spratt, 1990). Further, the way disease affects the dynamics of vertebrate populations is largely ecological and not a mere interplay between the host and the agent. Even if the pathogen is found very successful, studies should be made on the species specificity, impact on other organisms and most importantly safety to human beings. Parasites and pathogens might carry risks for non-target species, including man (Wood, 1985). Diseases which affect man which are known to be derived from wildlife are many. For example, rabies, Western encephalitis, St. Louis encephalitis, Colorado tick fever, rocky mountain spotted fever, relapsing fever, Q-fever, plague, tularemia, murine typhus, lymphocytic choriomeningitis, psittacosis, leptospirosis, salmonellosis, toxoplasmosis, etc. A variety of bacterial infections may be contracted from wild animals, notably those caused by *Pasteurella pseudotuberculosis*, *Pasteurella multocida*, *Bacillus anthracis*, *Erysipelothrix rhusiopathiae*, *Clostridium tetani* and *Listerella monocytogenes*. Certain of the fungus diseases, such as coccidioidomycosis and histoplasmosis are derived from exposure to wildlife habitats (Howard, 1967).

Chapter 10

Resistant or Tolerant Crops

The most common and important approach to animal and bird pest problem is to select and grow crops or varieties not highly prone to mammal or bird damage (*i.e.*, resistant crops). An approach that has not received much attention warrants the selection of parent stock with resistant characteristics and selective breeding to develop strains, hybrids or cultivars with improved resistance.

Animal Resistant Plants

Standard size apple trees suffer less damage than dwarf or semi-dwarf apple trees even in serious deer problem areas, because a greater proportion of the flowering buds are above the deer's reach (Moen, 1983). The same, of course, is true for other kinds of dwarf fruit trees. Clements (1980) pointed out that with sugarcane cultivars, the persistence of the leaves including sheaths influences the amount of rat (*Rattus* spp.) damage suffered. The self-stripping type canes are much more susceptible to rat damage than the so-called trashy ones. Varietal characteristics of rat resistant sugarcane are thick-barreled, moderate to hard rinded, grow erect and are not prone to lodging (King *et al.*, 1965). A land race of wheat having long (7.03 cm) and spreading awns and small and soft grains grown in Indian Himalayas is reported as not preferred by the monkeys, *Macaca mulatta* (Gupta *et al.*, 2009). Various chemical components (*e.g.*, phenols, terpenes, resins, essential oils) of tree species have been studied to determine the relationship between secondary compounds and resistance to animal damage with both positive and negative results (Oh *et al.*, 1967; Tucker *et al.*, 1976; Radwan and Crouch, 1978; Radwan *et al.*, 1982; Welch *et al.*, 1983; Tahvanainen *et al.*, 1985; Hansson *et al.*, 1986; Roy and Bergeron 1990).

The phenol compounds of coniferous trees play a significant role in deterring meadow voles, *Microtus pennsylvanicus* from debarking, which suggests that some selective breeding may be rewarding in preventing vole depredation (Roy and Bergeron, 1990). Promising lines were developed by breeding birch, *Betula* spp. for resistance against field vole, *M. agrestis* (Rousi *et al.*, 1988). Meadow vole resistant Norway spruce and pine seedlings were found to have 3-Myrcene and bornyl acetate apart from higher levels of limonene (Bucyanayandi *et al.*, 1990). 'Novole' is a crab apple which is reported to be highly resistant to the gnawing of the pine vole, *M. pinetorum* and meadow vole, *M. pennsylvanicus* (Cummins and Byers, 1984). The Chinese fir, *Cunninghamia lanceolata* with high resin content in the bark are found rejected by the red-bellied tree squirrel, *Calloscirus erythaeus* (Hwang *et al.*, 1985). While many farmers know from experience that pest mammals cause damage to specific crops, unfortunately very little has been published on which varieties, strains or cultivars should be selected to avoid the more serious mammal damage problems.

Bird Resistant Plants

Bird repellents that are effective have tended to be toxic, while those that are relatively nontoxic have tended to be ineffective. Thus interest has focused on the natural chemical defenses used by plants to defend themselves from herbivores. The use of biochemical or morphological genetic traits in a crop to protect ripening seeds or grain or fruit from bird damage remains a promising tool under certain situations. Bird resistance involves breeding varieties which are unattractive to birds and which are attacked only as a last resort. The search for bird-resistant varieties of cereal grains is likely to yield long-lasting results, even though early results have been somewhat disappointing (Howard, 1967). Chemical traits for this character generally involve an unpleasant taste, while morphological traits are usually those that impair feeding efficiency.

Depending on the crop, morphological or biochemical factors can predominate; in combination they can be synergistic (Harris, 1969). Pea varieties which are aphyllous and with more tendrils are found to discourage bird damage in the field. Animals and birds are discouraged because of the innate fear caused by the movement of a group of interlocked plants while feeding, since tendrils form a network with other plants. However, conditional aversion if created on the pest species is of more value than wading them after sampling.

Sorghum

Among the various morphological characteristics some bird resistance has been attained in genotypes with pendant heads, long awns, large glumes and large seeds (Doggett, 1957) that physically make bird feeding difficult. Tipton *et al.* (1970) observed that open-headed sorghum varieties were less damaged than compact ones. Open-head or lax-panicle trait has best inhibited perch feeding for birds larger than 50g. Common sparrows (*P. domesticus*), doves (*Stretopelia senegalensis* and *S. capicola*) preferred the compact-headed varieties over open headed ones. Perumal and Subramanian (1973) made similar observations at Tamil Nadu, India, where doves and parakeets (*P. krameri*) attacked more on compact headed sorghum. Large glumed cultivars were less damaged than that of short glumes because these glumes apparently make feeding more difficult for birds (Perumal and Subramanian, 1973).

Photo 16: Wheat landrace 'Tank' having large awns discourage animal.

Photo 17: Aphyllus pea having tendrils discourage bird or animal feeding/ entry.

Awned lemmas, which protrude beyond the tip of the grain, appear to hinder feeding activity of birds. Strong awned sorghum is more resistant to bird attack than awnless (Jowett, 1967; Perumal and Subramanian, 1973). The same is true for bajra also (Beri *et al.*, 1969).

Three blackbird-resistant varieties of grain sorgum (Northrup King 120, Northrup King 125, and Adkins-Phelps 614), although grown alongside heavily damaged corn field reported to have less than 1% damage (DeGrazio, 1964).

Tannins are the best known chemical components associated with bird resistance in agricultural crops. The tannin in these varieties gives an astringent taste/ tactile response in birds thereby inducing avoidance. In sorghums, this trait is stronger than all of the morphological traits; however, even high-tannin sorghums will be attacked if feeding pressure is severe (Doggett, 1970). The commercially available tannin containing hybrids 'brown sorghums' are also low palatable and also have nutritional problems. But several varieties reported resistant to bird depredation are with good nutritional quality in feeding or digestion trials also (Damron *et al.*, 1968; Harris *et al.*, 1970; Mabbayad and Tipton, 1975; Nelson *et al.*, 1975; Bullard, 1979). Other chemical components which repel birds are needed to be studied.

Corn

Ripening corn is subject to bird damage from the milk stage through harvest (Mitchell and Linehan, 1967). Most birds peck the center of immature kernels and remove the soft contents. Bird-resistant varieties generally had longer, heavier husks that were difficult for birds to penetrate. The size of the damaging bird in relation to husk characteristics is an important consideration. Larger birds are a more serious problem in corn than smaller birds. For example, the village weaver, *Ploceus cucullatus* and chestnut weaver, *Ploceus rubiginosus* have heavy, stout bills which can tear the husks and inflict damage (Bruggers, 1980) while the quelea, *Quelea quelea* and golden sparrows, *Passer luteus* are smaller and unable to open the husks (Erickson, 1979). Birds especially the blackbirds choose corn varieties on the basis of mechanical factors associated with the husk rather than on the basis of taste (Dolbeer *et al.*, 1982; Mason *et al.*, 1984). A tight-husked hybrid variety of corn that was thought to be bird-resistant, however, sustained heavier damage than surrounding corn because it was slow in maturing, being still in the vulnerable dough stage after the other varieties had begun to harden (DeGrazio, 1964). Long and heavy husked sweet corn is resistant to starlings, *S, vulgaris* (Dolbeer *et al.*, 1986b).

Rice

All of the bird-resistant rice research reported thus far involves morphological characteristics. Since lodging is the factor that makes this crop more susceptible to bird-damage, greater effort should be extended into testing or developing more upright varieties less prone to lodging. The flag leaf of rice can obstruct feeding access to the panicle, at least for small birds. Munias (*Lonchura striata* and *L. punctulata*) and weavers (*Ploceus philippinus*) were reported to cause less damage to varieties with erect flag leaves (Avery, 1979). In India, Uthamasamy *et al.* (1982) observed that the cultivars having a narrow angle between the flag leaf and panicle or a long flag leaf were damaged less by rose-ringed parakeets (*P. krameri*), sparrows and weavers. Rice awns also apparently reduced bird damage to ripening rice (Abifarin, 1984).

Sunflower

Chaff length, head angle, plant height and stem angle were important morphological factors involved in resistance to damage by sparrows, house finches and red-winged blackbirds (Seiler and Rogers, 1987). Concave and down-turned heads are aversive to sparrows (Parfitt, 1984). Several morphological traits have been observed to discourage red-winged blackbird predation: large, down-turned heads, strongly concave heads, stem curvature for which head to stem distance exceeds 15 cm, long involucre bracts, white seed (Parfitt, 1984) and involucre bracts which wrap around the maturing heads (Deodikar *et al.*, 1978). Long bracts which wrap around the face of the head and seed held tightly within the developing head are all features which contribute to bird resistance in sunflower. Purple-hulled genotypes have been reported to suffer less bird damage than oilseed types, possibly through taste aversion to anthocyanin pigment (Fox and Linz, 1983; Mason *et al.*, 1986). Harada (1977) reported that the chlorogenic acid in sunflower seeds and hulls may deter bird predation. However increased levels of chlorogenic acid in hulls or seeds did not deter feeding by blackbirds or sparrows (Parfitt *et al.*, 1986; Parfitt and Fox, 1986). Some of the above traits have been reported to be heritable (Fick, 1978) and thus chances for breeding a resistant strain.

Fruits

Bullfinches, *Pyrrhula pyrrhula* were found to favour pear varieties with low chlorogenic acid, a cinnamic acid derivative which belong to the group of plant secondary compounds known as the phenolics. The tannins also belong to this group and have long been implicated in plant defense (Rosenthal and Janzen, 1979). Citrus terpenes are reported to repel birds (Werner *et al.*, 2008). American robins, *T. migratorius* and European starlings, *S. vulgaris* avoid fruits with much sucrose because of the lack of digestive enzyme sucrase (Karasov and Levey, 1990). So fruit cultivars with as much as 15% sucrose is needed to repel robins and such high sugar concentrations naturally occur in fruits. Development of high sucrose cultivars would require almost complete replacement of simple sugars by sucrose and might be possible through selective breeding (Brugger, 1992). Humans find sucrose palatable and digestible and prefer its taste to that of glucose or fructose (Vettorazzi and MacDonald, 1988). Sucrose is a natural constituent of fruits and thus offers a nonlethal means for repelling birds in orchards and fields.

Conclusion

The use of effective bird-resistant varieties of crops, while not at present a reality, may yet prove to be one of the more promising means of combating bird depredations in the future (DeGrazio, 1964). This crop protection method involves feeding behaviour of granivorous birds and its effectiveness depends on the availability of preferred alternate foods. That is, bird-resistant traits provide protection to the crop when other food choices are readily available; however, when alternate food is scarce or high bird populations create serious feeding competition, they are less effective. Several practical factors like efficacy expectations, agronomic considerations and cost effectiveness need to be considered. In fields where most of the damage is caused by small birds, late in grain development, grain size and hardness are to be considered. Doggett (1957) observed that some grains are too large for the beak gap of small birds and they have difficulty in consuming them. Small birds may switch to smaller grass seeds if they are readily available. Results from preference studies indicated that kernel size had little effect on preference for sorghum grain during milk or dough stages, but the smaller grain was highly preferred over the larger in the ripened stage (Bullard and York, 1985). It is important that the bird-resistant cultivar must be competitive with high-yielding varieties grown locally. A farmer who selects a more damage-resistant cultivar of a crop may benefit; however, if all the farmers in the area do the same, it is very possible that no one will gain and the economic losses overall will remain about the same. Alternative food availability, in amount and variety, as well as preferences, play interrelated roles in animal damage problems and represent a basic principle that must be considered with the use of more resistant crops or plants. Preference of a plant varies with season and growth phase of plant and also the nutritional needs of the pest. Use of bird resistant cultivar can also be beneficial to integrated pest management programmes. Many birds feeding on crops also feed extensively on insects and if resistant cultivars are used these feeding habits of birds on insects may get enhanced (Dolbeer *et al.*, 1988).

Chapter 11

Regulatory Measures

11.1 Laws on Wildlife in India

The wildlife laws in India have a very long history. The earliest law is traced back to the third century BC, when Emperor Ashoka of Maghadha enacted a law to preserve wildlife and environment (Joshi *et al.*, 1998). The first codified law, The Wild Bird Protection Act, 1887 and the later Wild Birds and Animals Protection Act, 1912 were enacted by the British. Soon after Indian independence, the Indian Board of Wildlife (IBWL) was constituted in the year 1952 which took the task of preserving the natural wildlife habitats and saving the animals from probable extinction. After the U.N. Conference of Human Environment in Stockholm, the Indian Parliament has passed a comprehensive national law, The Wildlife (Protection) Act, 1972 for protection of plants and animal species. It extends to the whole of India, except the State of Jammu and Kashmir which has its own wildlife act. The Act was amended subsequently in 1982, 1986, 1991, 1993, 2002 and 2006 to accommodate provision for its effective implementation. It not only prohibits hunting but also deals with creating protected areas and controls trade in wildlife products. The foremost purpose of this act is to protect the habits of wild animals. As a development of this, various national parks and game sanctuaries have been established to ensure greater protection to wildlife.

11.2 The Wildlife (Protection) Act

The main object of the Act is to provide protection to the wild animals birds and plants. The Act empowers the Central Government to declare certain areas as Sanctuaries or National Parks. The Act prohibits hunting of wild animals; birds etc. and impose punishment for violating the same. The Act contains 66 Sections divided into seven chapters and six schedules. Chapter - I (Sec. 1 and 2) contains short title and definitions.

Chapter - II deals with Authorities under the Act. Chapter - III deals with the protection of Specified Plants. Chapter - IV provides for declaration of sanctuaries, National Parks and Closed Areas. Chapter - IV - A deals with Central Zoo Authority and Recognition of Zoos. Chapter - V deals with Trade or Commerce in Wild Animals, Animal Articles and Trophies. Chapter - V- A deals with prohibition of Trade or Commerce in Trophies, Animal Articles etc. Chapter - VI relates to Prevention and Detection of offences and finally Chapter - VII contains Miscellaneous Provisions. Having six schedules with varying degrees of protection, the act provides protection of wild animals, birds and plants. Schedule I and part II of Schedule II provide absolute protection - offences under these are prescribed the highest penalties. Species listed in Schedule III and Schedule IV is also protected, but the penalties are much lower. Schedule V includes the animals which may be hunted. Though hunting is not recommended, the chief wildlife warden has authorized to grant permission to hunt animals under certain circumstances.

1.   If an animal listed in schedule I has become i) dangerous to human life; or ii) is disabled; or iii) diseased as to be beyond recovery.

2.   If an animal enlisted in schedule II, III or IV, has become dangerous to human life or to property including crop or it is disabled or diseased beyond recovery.

(The Wildlife Protection Act, 1972; 2005)

Animals included in schedule I

Part I: Mammals

Andaman wild pig, Bharal/ blue wild sheep, Bearcat/ Binturong, Blackbuck, Brow-antlered deer or Thamin, Himalayan brown bear, Capped langur, Caracal, Whales, Dolphins and Porpoises, Cheetah, Chinese pangolin, Chinkara or India gazelle, Clouded leopard, Crab-eating macaque, Desert cat/ African wild cat, Desert fox, Dugong, Ermine/ stoat, Fishing cat, Four-horned antelope/ Chausingha, Gangetic dolphin, Gaur or Indian bison, Golden (Asian) cat, Golden langur, Giant (grizzled) squirrel, Himalayan ibex, Himalayan tahr, Hispid hare, Hog badger, Hoolock gibbon, Indian elephant, Indian (Asiatic) lion, Indian wild ass, Indian (grey) wolf, Kashmir stag/ Hangul, Leaf monkey, Leopard or Panther, Leopard cat, Lesser or red panda, Lion-tailed macaque, Loris (slender loris), Little Indian porpoise, Lynx, Malabar (large spotted) civet, Malay or Sun Bear, Marbled cat, Markhor, Mouse deer (chevrotain), Musk deer (Siberian), Nilgiri langur, Nilgiri tahr, Nyan or great Tibetan sheep, Pallas's Cat, Pangolin (Indian) , Pygmy hog, Ratel/ Honey badger, Rhinoceros, Rusty spotted cat, Serow, Clawless otter, Sloth bear, Slow loris, Small Travencore flying squirrel, Snow leopard, Snubfin dolphin, Spotted linsang, Swamp deer, Takin or Mishmi takin, Tibetan antelope or Chiru, Tibetan fox, Tibetan gazelle, Tibetan wild ass, Tiger, Urial or Shapu, Wild/ water buffalo, Wild yak, Tibetan wolf, Wroughton's free tailed bat and Salim Ali's fruit bat.

Part II: Birds

Andaman teal, Assam bamboo partridge, Bazas, Bengal florican, Black-necked crane, Blood pheasants, Cheer pheasant, Eastern (Oriental) white stork, Forest-spotted owlet, Frogmouths, Great Indian bustard, Great Indian hornbill, Hawks, Hooded crane, Hornbills, Houbara bustard, Hume's bar-backed pheasant, Indian pied hornbill,

Jerdon's courser, Lammergeier, Large falcons, Large whistling teal, Lesser florican, Monal pheasants, Mountain quail, Narcondam hornbill, Nicobar megapode, Nicobar pigeon, Osprey or fish-eating eagle, Peacock pheasants, Peafowl, Pink-headed duck, Sclater's monal, Siberian white crane, Tibetan snow-cock, Tragopan pheasants, White-bellied sea eagle, White-eared pheasant, White/ Eurasian Spoonbill, White-winged wood duck, Swiftlets, Hill myna, Tibetan eared pheasant, Kalij pheasant, Lord Derby's parakeet, Vultures and White bellied herons.

Apart form Wildlife Protection Act, many laws, like Indian Forest Act, 1927; Environment (Protection) Act, 1986; Prevention of Cruelty to Animals Act, 1960; Biological Diversity Act, 2002 were enacted to protect wildlife and control illegal trade (Singhar, 2002). Further, India is also a party to Convention on International Trade in Endangered Species of Wild Fauna and Flora (CITES). Despite all these laws and policies, conflict between wildlife and human still exists. Monkey, deer, wild boar and elephant destruction of crops has become very common. Human and domestic animal death due to attacks of wide ranging large animals also occur.

However, vertebrate pest management should take into consideration on the status of the organism in the ecosystem also. The conservation status of an organism indicates whether it is extant and how likely the group is to become extinct in the near future. The IUCN Red List of Threatened Species is the best-known worldwide conservation status listing and ranking system.

- Extinct (EX) – No known individuals remaining.
- Extinct in the Wild (EW) – Known only to survive in captivity, or as a naturalized population outside its historic range.
- Critically Endangered (CR) – Extremely high risk of extinction in the wild.
- Endangered (EN) – High risk of extinction in the wild.
- Vulnerable (VU) – High risk of endangerment in the wild.
- Near Threatened (NT) – Likely to become endangered in the near future.
- Least Concern (LC) – Lowest risk. Does not qualify for a higher risk category. Widespread and abundant.
- Data Deficient (DD) – Not enough data to make an assessment of its risk of extinction.
- Not Evaluated (NE) – Has not yet been evaluated in this criterion.

(http://en.wikipedia.org/wiki/Conservation_status)

So integrated pest management with biodiversity conservation is the need of the hour to frame a line which allows every living organism including man to live in harmony.

For further readings:

www.moef.nic.in/legis/wildlife/wildlife1.html

http://haryanaforest.gov.in/wildlifeprotectionact.aspx

Chapter 12

Integrated Vertebrate Pest Management

Classic philosophies of integrated pest management use all the available methods to suppress a pest population with minimum environmental impacts for the betterment of human interests (Fairaizl, 1992). However, Integrated vertebrate pest management (IVPM) is distinct from other IPM programmes and can be classified into four broad groupings such as multiple species management, management with multiple objectives/aims, management using different methods and at last the management of different species with different objectives, technologies and strategies (Coleman, 1993).

12.1. Multiple-species management

This is termed as simultaneous management of two or more pest species which is normally achieved by broad-spectrum baiting. For instance, red deer and possum management using toxic carrot bait and possum and rat management using toxic cereal baits. Poisoning with pindone controlled the target rabbit population and also mice and thus an integrated management of both with one methodology (Parkes, 1996).

12.2. Multiple-objective management

Management of one species of pest can be undertaken for two or more reasons. Pests affect the environment in diverse ways *viz.*, possums are vectors and reservoirs of bovine Tb and has a negative impact on botanical and wildlife values on the conservation estate. Possums also damage catchment plantings, fodder crops, exotic plantations and pasture production. Likewise, rodents are pest of crops, storage and domestic holdings besides vector of serious diseases.

12.3. Management using multiple technologies and strategies

Integrated pest management is an approach that involves the use of physical, biological, chemical and also cultural practices. Any single method will not be a panacea for vertebrate pest control needs, but they will probably become a valuable adjunct to integrated control of pest species. Integration of multiple control techniques is likely to be more effective than using individual techniques (Inglis *et al.*, 1983; Mason *et al.*, 1989; Dolbeer 1990). A combination of efficient methods will probably make synergistic effects possible, with the degree of control greatly exceeding the sum of the independent effects of each method.

12.4. Fully integrated management

Management that integrates different species, objectives, technologies and strategies is the ideal for mammal and bird pest management. Once efficacies of individual techniques are established, a second suite of experiments that integrates multiple techniques to assess their efficacy in combination will be required. Finally and most importantly greater emphasis needs to be placed on the underlying ecological principles that are associated with the desired animal responses to deterrent techniques. Once understood, the ecological principles that have previously resulted in the limited effectiveness of harassment, deterrent and repellent techniques can be used to modify these techniques and maximize their effectiveness. Most importantly these integrated techniques should be used against the real pest problem which causes multiple damages to the ecosystem. Further, integrated management of vertebrate pests cannot be achieved by simple integration of many methodologies for a noxious pest but only through an ecological approach through the following steps

    i.   Influencing food and shelter to reduce pest populations or damage

    ii.   Increasing natural predation rates

   iii. Exclusion of pest animal from areas to be protected

   iv. Integrated vertebrate pest management (IVPM)

I. Influencing food and shelter to reduce pest populations or damage

Cultural control practices like minimizing seed spillages and fly control in feedlots should be added, apart from bird control strategies in IPM programmes to manage pest birds in feedlots (Palmer, 1976). The growth and maturation of weeds, grasses and seeds after rainfall in agricultural fields provides rats with a source of protein required for reproduction. Therefore, weed control within and around crops also reduces nutritional support for breeding and thus reduction in pest population (Smith *et al.*, 2002). Sanitation is the best method of rat control in households. Grain storage should be done in tight, rat proof containers so as to avoid the accessibility of rodents to the food. Clearing of vegetations has reported to increase the vulnerability of rabbit and rodent predation by fox and owl in pine plantations (Munoz and Murua, 1990). In agricultural areas, eliminating nest-sites and refuge areas is also vital and can be very effective in rodent management (Jackson and vanAarde, 2003). Maintenance of buffer zones and wildlife boarders are the most effective and sustainable management option for crop field adjacent to woody mountain and forest areas.

II. Increasing natural predation rates

Natural enemies maintain the pest populations in threshold levels, especially in an ideal ecosystem. When there is an ecosystem unbalance, often by artificial means or sometimes climate driven, which favours a particular organism, taking upper hands in the form of pest and destroys the ecosystem. Thus maintaining the natural predation rate is the best, eco-friendly and sustainable management practice of vertebrate pests. The combined predation pressure by native hawks, owls, snakes and carnivores usually influences the seasonal and annual density of species of vertebrate prey than would otherwise exist. House cats were found to control mice and rats and also resident house sparrow up to 80%, especially when retained in particular area and periodically fed with some supplemental food. Artificial perches were reported significantly increased raptor numbers in agricultural areas (Hall *et al.*, 1981). Artificial perches and nest boxes were placed in orchards to attract birds of prey to reduce vole populations (Askham, 1990). Increasing predator abundance and efficacy through cultural practices is explained by Munoz and Murua (1990), wherein the effect of predators on prey was studied by clearing 4 m wide strips of all vegetation and providing perches for predatory birds. The modification of the habitat attracted the barn owl, *Tyto alba* and improved visibility and potential for aerial maneuvering, increased the visibility and ease of movement of the South American fox, *Pseudalopax* sp., and increased the vulnerability of rodents and rabbits to predation.

III. Exclusion of pest animal from areas to be protected

Exclusion of a particular pest from areas need protection (crop field/ storage) is more effective and economical than attracting and killing. Fencing or netting is the best way of complete exclusion of the pests from the area of concern. Fencing when made properly may exclude a variety of mammal pests *viz.*, burrowing rodents to pushing elephants. For this reason only, electric fencing is becoming more popular especially for bigger mammals management. Bird netting is the only full proof technology to prevent bird damage especially in high value orchard crops and aquaculture facilities. Rat proofing is the best management option available for rodent pests in storage godowns which includes high plinth, 25 cm metal sheet lining at the door bottom and netting of windows and ventilators. In agricultural sites also rat are reported to take only 2 to 8 weeks to recover even after 70% are killed or removed (Lu *et al.*, 1994). So deterrence or proofing is much effective than mortality, trapping or control.

IV. Integrated vertebrate pest management (IVPM)

In some situations, pest populations may increase above economic threshold levels even when all the techniques dealt above, such as reducing food and shelter; increasing predation, exclusion etc. are well practiced. In such cases, management of pests to reduce the damage is necessary and there are many management options which are discussed in earlier chapters. A well planned IPM program for vertebrate pests includes identification of the problem, determining acceptance thresholds, defining precise goals and objectives and developing and implementing a monitoring program (Coffey and Johnston, 1997; Engeman and Witmer, 2000). There are several examples of integrated approaches having improved effectiveness over single techniques (Montoney and Boggs, 1995; Belant, 1997; Tobin, 1998).

12.5 Pest Management through Different Methods

Different sounds

A combination of two or more different sounds is often needed to move the birds out of the crop (Beason, 2004) especially from a well established feeding spot. Frightening devices such as carbide and acetylene exploders, exploding shotgun shell, rope firecrackers, shotgun and the .22 rifle, when used to supplement one another is reported very effective in preventing bird damage (DeGrazio, 1964; Hobbs and Leon, 1987). Bird bangers, screamer sirens, shell crackers and propane cannons emit loud noises which emulate a shotgun blast or loud whistle. When used in conjunction with shooting, these techniques can be very effective. The most effective method of scaring blackbirds from rice fields has proven to be a combination of propane exploders, pyrotechnics and biosonics. A combination of distress calls and pyrotechnics was most effective than any one alone and both in combination was used in many situations effectively for bird dispersals (Brough, 1965; Brough, 1968; Busnel and Giban, 1968; Bridgman, 1969; 1976; Dahl, 1984). Mott and Timbrook (1988) demonstrated greater reductions in goose abundance at campgrounds when incorporating goose alarm and distress calls with pyrotechnics. Brough (1967) performed a 12 month study scaring gulls, corvids, lapwings and starlings using warning calls and shell crackers. When shell crackers were used alone they were successful on 88 of 120 occasions (73%). When broadcast calls were used alone they were successful on 476 of 573 occasions (83%). Finally, when broadcast calls and shell crackers were used in combination they were successful on 285 of 306 occasions (93%). This research showed the effectiveness of broadcast warning calls as well as the increased effectiveness of calls used in conjunction with a secondary scaring method.

Visual and Auditory

Visual and auditory are two important sensory repellents when used in combination to deter a bird or animal pest have a synergistic effect. The combination of sound and visual scare device works much as bird deterrents better than either by itself (Pritts, 2001). The reflective ribbon itself imitates fire with a flashing effect and also produces a humming or crackling noise when moved by the wind. An eye-spotted ball is found to significantly reduce the number of birds visiting a bird table to feed. Deterrence was increased 10% by reinforcing the visual stimulus with alarm calls of common blackbirds, house sparrows and starlings (McLennan *et al.*, 1995). In most situations, traditional scarecrows and models of perched raptors do not closely enough resemble a situation that is alarming or threatening to birds (Inglis, 1980). Field studies have indicated that mechanically incorporating sound stimuli into the models may greatly enhance their effectiveness (Zajanc, 1962; DeHaven, 1971; Pierce, 1972; Bivings, 1986). Scarecrows are considered to be of little value in deterring blackbirds from rice fields unless they are used with pyrotechnics or exploders (Neff and Meanley, 1957; DeHaven, 1971). Ward (1978) tested a large fluorescent orange scarecrow, two lights and a gas cannon and found very effective against ducks, in particular lesser scaup, *Aythya affinis*. A human-like scarecrow that popped up from a double propane cannon when it fired was reported very successful in keeping blackbirds from feeding over 4-6 acres of sunflowers (Nomsen, 1989). If distress call playbacks are paired with a predator model, especially one that appear to be grasping the caller, birds that approach the sound source might have their initial fears reinforced rather than get lightened (Conover, 1994) and its efficiency can be further enhanced by tying it with a live bird. Howard *et al.* (1985) suggested designing models that display action or produce sound, which is somehow triggered by the pest birds when they first enter an area, before they have a chance to land and feed. Such action or sound should be discontinued when the birds leave and thus make less rapidly habituation.

Animal induced pop-up scare crow coupled with propane exploders are found to reduce blackbird damage by 71 to 87% in ripening sunflower (Cummings *et al.*, 1986). Reinforcement of human effigies or predator models with shooting is highly recommended to increase their effectiveness. The use of Scary Man® in combination with harassment patrols *i.e.* shooters with 12 guage shotgun shell and 0.22 caliber rifles drastically reduced the cormorants in aquaculture tanks (Stickley and King, 1993).

Villagers in Sumatra use powerful flashlights in combination with noise and fire to deter crop raiding elephants with some success (Grant *et al.*, 2007). But strobe lights with sonic acoustic devices did not reduce deer use of sites nor alter their movements. Motion activated light and sound emitting frightening device proved ineffective for elk, mule deer and white-tailed deer (VerCauteren *et al.*, 2005b).

Tactile and Auditory

Broadcasting of sounds (8 alarm calls) along with surface modification like pasting high density polyethylene sheets in bridges to deter cliff swallows from perching and nesting was found to reduce by 82.8% nesting. The polyethylene sheets alone and broadcast calls alone were found to reduce the same by 69.1% and 53.0%, respectively (Conklin *et al.*, 2009).

Chemical and Visual

Methyl anthranilate, a chemical repellent was reported to be effective against snow goose, *Chen caerulescens* for about one week period but when mixed with a white paint pigment (titanium oxide, TiO_3) enhanced the repellency up to 21 days (Mason and Clark, 1996). Titanium oxide and calcium carbonate are also reported as visual cue additives to aerially applied methiocarb formulations for bird deterrence. In fact, simple application of calcium carbonate is not noticed to deter the geese (Dolbeer *et al.*, 1992). Coloured plastic flags which itself a repellent (Mason and Clark, 1994) is also found to enhance methyl anthranilate repellency to goose. Waterfowl use of ponds was reported to reduce significantly when motion-activated frightening devices were integrated with a chemical repellent (Stevens *et al.*, 2000).

Biological and Chemical

The successful integrated management of European rabbits, *O. cuniculus* in Cabbage tree Island, Australia involved the serial introduction of myxoma virus, haemorrhagic virus and finally poison baiting programme (Priddel *et al.*, 2000).

Mechanical exclusion and Chemical/ Biological

Apiaries baited with emetic chemical, lithium chloride in bee boxes and kept outside the fence at the likely avenues of bear approach have reported to reduce black bear damage significantly (Gilbert and Roy, 1977). Two guard dogs within a simple fence are found very effective for protecting 2 ha area of white pine, *Pinus strobus* seedlings and ornamentals from deer damage (Beringer *et al.*, 1994).

Electrical and Olfactory

Combining electric fences with attractants may encourage the animal to touch the fence with their nose or mouth, thereby enhancing aversive conditioning behaviour. Early studies by Kinsey (1976) and Porter (1983) used aluminum flags coated with peanut butter to attract deer to an electrified, single-strand smooth wire. This design was reported to be effective for sites less than 5 ha, with light to moderate foraging pressure by deer. Hygnstrom and Craven (1988) used fences constructed from an electrified ribbon and treated the entire length with a peanut butter-oil mixture and found effective. However, using an attractant to attract and give a conditional aversion (shock) or using a repellent to add the effect of electric shock is a matter of debate. Jordan and Richmond (1992) evaluated the relative effectiveness of attractants *vs.* repellents for excluding deer with a three-wire electric fence system. The electric fence with a repellent was penetrated by deer only once whereas it was nine times in electric fence with peanut butter and thirteen times in control, thus the electric fence with repellents can be concluded as the most effective barrier. An electric fence treated with predator urine may significantly reduce deer damage (Hygnstrom and Craven, 1988). A single-strand, electrified, plastic ribbon fence treated with bobcat, *Lynx rufus* urine was found to be very effective in preventing woodchuck, *Marmota monax* damage to a cabbage field (Curtis and Petzoldt, 1994).

More Combinations

Integrated hazing approach using pyrotechnics, dogs and lasers has become a popular method of deterring geese from a site (Castelli and Sleggs, 2000; Swift, 2000; York *et al.*, 2000). Holevinski *et al.* (2007) found that a remote controlled boat and guard dog, Border collie combination removed greater than 90% of geese. Broadcasting of distress calls of starlings, finches and robins added with conventional methods like reflective tape, propane cannons and pyrotechnics significantly reduced damage when compared with only conventional control (5.7% *vs.* 13.0%) in vineyards (Berge *et al.*, 2007). A combination of lasers, distress calls and pyrotechnics was 98% effective in reducing crow abundance at urban roosts (Chipman *et al.*, 2008). Lethal control can also enhance the efficacy of harassment and deterrent techniques of birds at airports (Dolbeer *et al.*, 1993).

An IPM strategy of trapping or poisoning and using tree guards is recommended to reduce damage by animal browsing in new tree plantations (Walsh and Wardlaw, 2005). For preventing elephant raids on crops, in Zimbabwe, cheap farmer-based traditional methods have been shown to be very successful when used in combination. This includes clearing of 5 m boundaries around fields; strategically placed watchtowers, use of whistles by the guards; cowbells on string fences as a vigilant system. Burning of fires on field boundaries and at known elephant entry points; 'brickettes' of dried elephant dung and ground chillies being burnt to create noxious smoke; a mixture of chilli pepper oil and grease being smeared on the fences and planting of chillies as an unpalatable buffer crop along with use of whips to imitate gunfire; throwing firecrackers at the elephants (Hoare, 2001a). Maximum deterrence can be achieved by the combination of the greatest number of methods.

Table 12.1: Integrated Animal and Bird Pest Management Tactics

Pest	Frightening/ Management techniques	Comments
Blackbirds	Gas exploders, scare crow/predator effigies, pyrotechnics, mylar ribbon, chemicalfrightening agents and toxicants	Habituation limits effectiveness, help to move flocks to other areas, Integrated approach may improve results
Parrots	Mylar ribbon, bagging, trapping, human p atrol	Reflective ribbon and human patrol are effective
Geese	Gas exploders, mylar ribbon, flags, reflective objects, distress/alarm calls	Habituation may limit effectiveness
Gulls	Pyrotechnics, mylar ribbon distress/alarm calls, shooting, chemical frightening agents and toxicants	Habituation may limit effectiveness, integrated approaches may improve results
Picivorous birds	Gas exploder, mylar ribbon, pyrotechnics, lasers, scare crow/ predator effigies, alarm/ distress calls	Effectiveness is variable, integrated approach may improve results
Pigeons	Trapping, human patrol, chemical frightening agents and toxicants	Trapping may reduce the population
Starlings	Distress/alarm calls, predator effigies, chemical frightening agents and toxicants	May help move roosting/feeding flocks to other areas
Crows	Distress calls, trapping, poisoning	Australian crow trap is reported very effective
Passerines/ song birds	Flags, exploders, repellent chemicals	Some are predatory (beneficial) and many have high aesthetic value so care should be taken for management
Elephants	Fire, drums, capsicum oleoresin, human patrol, sturdy electric fencing	More combination of methods leads to more success
Bears	Trapping and relocation, electric fencing	Sturdy and high electric fencing for apiaries
Deer, antelopes, blackbucks	Gas exploders, rope firecrackers, revolving lights, guard dogs, fencing	May help move migrating herds on to other areas
Pigs	Trapping, poisoning, shooting, fencing	Fencing and shooting are practiced extensively
Foxes	Poisoning, trapping, sterility agents, gas exploders, rope firecrackers, revolving lights	Sterility agents may be a good alternative to trapping or poisoning
Coyotes	Trapping, poisoning, aerial hunting, gas exploders, rope firecrackers, electronic guard	Highly unpredictable in their response to frightening devices
Monkeys	Guard dogs, sterility agents, human patrols	Trained guard dogs may be effective
Goats and donkeys	Fencing, shooting, poisoning	Fencing may the best
Rabbits	Diseases, predators, poisoning, gas exploders, firecrackers	Myxomastosis and haemorrhagic diseases are very effective. Can be used where commercial rearing is not practiced
Squirrels	Tree protectants, trapping, poisoning,	Arboreal traps are reported very effective
Raccoons	Lighting the area, playing a radio, gas exploders, pyrotechnics	Raccoons accustomed to people are difficult to frighten
Badgers	Flash light, trapping	Badgers are highly repulsive to light
Porcupines	Trapping, hunting, tree protectants, predator & other odour repellents	Odour repellants are reported very effective
Rodents	Exclusion, poisoning, trapping, using predators, diseases	Frightening techniques rarely have any appreciable effects on small rodents

Based in part on Koehler *et.al.* (1990) and Gilsdorf *et.al.* (2002)

Efficacy of different methods and their sequence

Selection of a method for vertebrate pest management should be based on its efficacy rather than easiness or quickness. The long term sustainability should be kept in mind in framing an IPM programme. For instance, the use of bird-resistant cultivars of corn, the deployment of frightening devices during the critical damage period and the provision of alternative feeding sites are important components of an integrated management programme on red-winged blackbirds (Dolbeer, 1990). IPM strategy of rodents in crop fields includes early detection of increasing rat populations, management of harbourage areas; weed control in and around crop fields and lastly by using toxicant baits. This combined approach reduces the potential for large scale rat plagues and reduces crop damage throughout the season (Smith *et al.*, 2002).

Time of management and threshold levels

Decision on the time to start management is very critical to solve the pest problem rather than the use of different techniques. This can be dealt in two ways *viz.*, critical crop stage and critical stage of the pest at which the management should be taken. For example, red-winged blackbirds, *A. phoeniceus* are often initially attracted to corn fields to feed on insect pests during the 2 to 3 week period between silking and kernel development. Once birds switch from feeding on insects to feeding on corn, protective measures to reduce damage are to be taken, more critically in the early (milk) stages of kernel development than later because the damage potential is much higher at earlier stage (Dolbeer, 1990). To arrive at this conclusion, accurate measurement of crop losses by a particular pest is to be evaluated, which is the missing link in many vertebrate pest management programmes. Rodent damage to rice was measured at several stages of crop growth by Singleton *et al.* (2005). In monoculture lowland rice, rodent damage to rice was 54% at the primordial stage, 32% at the booting stage, but only 16% at the ripening stage. But much of the damage assessments and management are being done at the ripening or harvest stages.

The use of threshold pest densities to initiate pest control can increase control effectiveness by reducing opportunity costs. However, complex trophic relationships between pests and resources make it difficult to set threshold levels. Vertebrate pest damages are often difficult to quantify and much more difficult to predict. Models on damage functions can be used to set threshold pest densities that achieve tactical but not strategic conservation outcomes. For instance, every 1% increase in tiller damage by rats was reported to cause a reduction of 58 kg/ha in rice yield (Singleton *et al.*, 2005). But how many rats can cause that 1% tiller damage and in how many days are need to be studied. Density dependent predator–prey models can be used to set threshold pest densities that have strategic consequences for resource conservation, but are limited in their scope where pest or resource abundance is influenced by density independent environmental perturbation (Choquenot and Parkes, 2001). Important components/ modules of a comprehensive decision support system for vertebrate pest management include an overview of the species biology and ecology, damage identification and monitoring methods, a description of damage potential and associated factors, a mechanism to evaluate alternative management techniques and the integration of techniques, a cost-benefit analysis component and sources of additional information (Witmer, 2007). Threshold level is based on the perception of the communities also because some people cannot tolerate pest damage but some, up to certain extent. It was reported that 36% of communities of Sahelian wetland perceived vertebrate pests as really pests whereas 28% of people as a source of food. Further, some communities interviewed said to tolerate 5% crop loss caused by vertebrate pests whereas some others up to 50% (Ezealor and Giles, 1997). Further, farmers discernment on crop damage caused by wild animals also vary heavily. All crop raiders do not cause same level of damage. Wild boars cause more damage to crop plants than monkeys even when these do frequent raids than the former (Newmark *et al.*, 1993). In some crops management decision has to be taken accordingly.

Cost effectiveness and environmental cost

Many a times, blanket vertebrate control would be uneconomical (Dolbeer, 1981; Foran *et al.*, 1985). So extensive surveys to identify areas where pest management can be carried out economically is to be done before initiation. Exclusion netting has been found to be highly effective and the best approach for orchards, vineyards and fish farms where it is economically viable (Bomford and Sinclair, 2002). Cost effectiveness can be predicted by modeling the damage potential of the pest and input costs. Cost effective pest management is achieved by starting the management operation when the pest crosses economic threshold levels. Thus IVPM is also an economic management of pests with effective technologies. Some cases, the benefit cost ratio will be very high especially for noxious pests such as rodents. The benefit-to-cost ratio of rodent control in rice averaged 25:1 but varied considerably and even reported as high as 63:1 in some situations. The economic benefit of integrated rodent management was equal to or better than that achieved by conventional management based on synthetic rodenticides (Singleton *et al.*, 2005). But when we look into the B:C ratio alone, poisoning may be of much economical rather than any other strategy. Coblentz and Baber (1987) reported poisoning as 11 times cheaper than shooting and 80 times cheaper than trapping of mammals. The environmental problems caused by toxicants especially of the persistent ones should also be taken into account, as environmental cost. Thus, IVPM aims at reduction of pest damages by using tolerant plants as a basis and other techniques like exclusion of pest or repelling them away and at the last by using toxicants. An additional concern, receiving more attention in recent years, is who should pay for the cost of vertebrate pest population and damage management activities that benefit the general public or the agriculturalists of a region? (Witmer, 2007). Vertebrate pests are more dynamic causing damage to a wide variety of habitats and not confined to an individual farm or area. Thus, awareness creation about the vertebrate pest damage and their alleviation methodologies to all the stakeholders is necessary.

System approach in IVPM

Many of the vertebrate pest management programmes appears to manage a single species. In some situations, a system approach of IVPM may be beneficial than a species management per se. A systems approach should integrate preventive measures to make an area or situation unsuitable or less attractive to all problem wildlife species that have been identified as economically or socially important. It should also incorporate control measures to remove problem individuals or to maintain populations at levels in which the damage they cause is economically and socially tolerable (Hygnstrom, 1990). The systems approach should integrate all the possible and useful methods, including habitat modification, exclusion, frightening devices, repellents, chemosterilants, toxicants, trapping, shooting and others. Care must be taken that control methods applied at the community level have minimal impacts on non-target species (Howard, 1976). The systems approach to wildlife pest management leads to impacts on a larger scale with long term cost effective and ecologically sound benefits than does the species management (Hygnstrom, 1990).

Integrated vertebrate pest management (IVPM) should be aimed at damage reduction rather than reducing the population.

References

Aars, J., X. Lambin, *et al.* 2001. *Anim. Conserv.*, **4**: 187-194.

Abifarin, A.O. 1984. *West African Rice Dev. Assoc. Tech. Newsl.*, **5**: 27-28.

Able, K.P. 1982. *Nature*, **299**: 550-551.

Achiron, M. 1968. *Natl. Wildl.*, **26**: 18-21.

Acorn, R.C and M.J. Dorrance. 1998. AGDEX 684-19. Alberta Agriculture, Food and Rural Development, Edmonton, Canada.

Agrawal, S., S. Chauhan and R. Mathur. 1986. *Andrologia*, **18**: 125-131.

Aguilera, E., R.L. Knight and J.L. Cummings. 1991. *Wildl. Soc. Bull.*, **19**: 32-35.

Ahmad, S., H.A. Khan, M. Javed and K.U. Rehman. 2012. *Int. J. Agric. Biol.*, **14**: 286-290.

Aitken, R.J and M. Paterson. 1988. *Nature*, **335**: 492-493.

Akande, M. 1978. *Proc. Vertebr. Pest Conf.*, **8**: 224-225.

Alexander, P.S., L.K. Lin and B.M. Huang. 1987. *J. Taiwan Museum*, **40**: 1-7.

Alford, B.T., R.L. Burkhart and W.P. Johnson. 1974. *J. Am. Vet. Med. Assoc.*, **164**: 702-705.

Alterio, N. 2000. *New Zealand J. Ecol.*, **24**: 3-9.

Alterio, N., K. Brown and H. Moller. 1997. *J. Zool.* (London), **243**: 863-869.

Amarante, C., N.H. Banks and S. Max. 2002. *New Zealand J. Crop Horti. Sci.*, **30**: 93-98.

Andelt, W.F., D.L. Baker and K.P. Burnham. 1992. *J. Wildl. Manage.*, **56**: 164-173.

Andelt, W.F., K.P. Burnham and J.A. Manning. 1991. *J. Wildl. Manage.*, **55**: 341-347.

Andelt, W.F., R.L. Phillips, K.S. Gruver and J.W. Guthrie. 1999. *Wildl. Soc. Bull.*, **27**: 12-18.

Andelt, W.F., T.P. Woolley and S.N. Hopper. 1997. *Wildl. Soc. Bull.*, **25**: 686-694.

Anderson, P.J., G.S.N. Lau, W.R.J. Taylor and J.A.J.H. Critchley. 1996. *Hum. Exp. Toxicol.*, **15**: 461-465.

Anderson, T.E. 1969. **In:** Wildlife management techniques. The Wildlife Society, pp. 497-520.

Anthony, R.M., V.G. Barnes Jr. and J. Evans. 1978. *Proc. Vertebr. Pest Conf.*, **8**: 138-144.

Arlet, M.E and F. Molleman. 2010. *African Primates*, **7**: 27-34.

Armistead, A.R., M. Ken and G.E. Connolly. 1994. *Proc. Vertebr. Pest Conf.*, **16**: 31-35.

Arnemo, J.M., T. Negard and N.E. Soli. 1994. *Rangifer*, **14**: 123-127.

Arnold, G.W., D.E. Steven and J.R. Weeldenburg. 1989. *Aust. Wildl. Res.*, **16**: 85-93.

Askham, L.R. 1990. *Proc. Vertebr. Pest Conf.*, **14**: 144-148.

Askham, L.R. 1992. *Proc. Vertebr. Pest Conf.*, **15**: 137-141.

Avery M.L and A.C. Genchi. 2004. *Wildl. Soc. Bull.*, **32**: 718-725.

Avery M.L., D.S. Eiselman, M.K. Young, J.S. Humphrey and D.G. Decker. 1999. *North Am. J. Aqua.*, **61**: 64-69.

Avery, M.L and D.G. Decker. 1994. *J. Wildl. Manage.*, **58**: 261-266

Avery, M.L and J.L. Cummings. 2003. *Proc. Symp Manag of North Am. Blackbirds*, The Wildlife Society Eighth Annual Conf., Bismarck, ND, pp. 41-48.

Avery, M.L. 1979. *Proc. Bird Control Sem.*, **8**: 161-166.

Avery, M.L., D. Decker and J.S. Humphrey. 1998. *Proc. Vertebr. Pest Conf.*, **18**: 354-358.

Avery, M.L., D.E. Daneke, D.G. Decker, P.W. Lefebvre, R.E. Matteson and C.O. Nelms. 1988. *Proc. Vertebr. Pest Conf.*, **13**: 277-280.

Avery, M.L., D.G. Decker and D.L. Fischer. 1994. *Crop Protec.*, **13**: 535-540.

Avery, M.L., D.G. Decker and M.O. Way. 1989. *Denver Wildl. Res. Cent. Bird Sec. Res. Rep. No.* **444**: 9.

Avery, M.L., D.G. Decker, J.S. Humphrey and C.C. Laukert. 1996. *Crop Protec.*, **15**: 461-464.

Avery, M.L., D.G. Decker, J.S. Humphrey, E. Aronov, S.D. Linscombe and M.O. Way. 1995. *J. Wildl. Manage.*, **59**: 50-52.

Avery, M.L., E.A. Tillman and C.C. Laukert. 2001. *Int. J. Pest Manage.*, **47**: 311-314.

Avery, M.L., J.S. Humphrey and D.G. Decker. 1997. *J. Wildl. Manage.*, **61**: 1359-1365.

Avery, M.L., J.S. Humphrey, E.A. Tillman, K.O. Phares and J.E. Hatcher. 2002. *J. Raptor Res.*, **36**: 44-49.

Avery, M.L., J.W. Nelson and M.A. Cone. 1992. *Proc. Eastern Wildl. Damage Control Conf.*, **5**: 105-110.

Avery, M.L., S.J. Werner, J.L. Cummings, J.S. Humphrey, M.P. Milleson, J.C. Carlson and T.M. Primus. 2005. *Crop Protec.*, **24**: 651-657.

Baer, N.W. 1980. *Tree Planters' Notes*, **31**: 6-8.

Bahr, J., R. Erwin, J. Green, J. Buckingham and H. Peel. 1992. Rep. from the Delta Environmental Management Group Ltd., and Southwest Research Institute, for Transport Canada.

Bailey, P.T and G. Smith. 1979. *Aust. J. Exp. Agric. Anim. Husb.*, **19**: 247-250.

Baines, D and M. Andrew. 2003. *Biol. Conser.*, **110**: 169-176.

Baker, S.J and C.N. Clarke. 1988. *J. Appl. Ecol.*, **25**: 41-8.

Ball, S.A. 2009. *Human–Wildlife Conflicts*, **3**: 257-259

Balser, D.S. 1964. *J. Wildl. Manage.*, **28**: 352-358.

Balser, D.S. 1965. *J. Wildl. Manage.*, **29**: 438-442.

Banks, P.B., K. Norrdahl, M. Nordstrom and E. Korpimaki. 2004. *Oikos,* **105**: 79-88.

Barden, M.E., S. Dennis, R.T. Calvert and D.W. Paul. 1993. *Proc. East. Wildl. Damage Control Conf.*, **6**: 22-29

Barnett, A and J. Dutton. 1995. Geography Outdoors, Royal Geographical Society with IBG, London. 131 p.

Barreto, G.W., S.P. Rushton, R. Strachan, *et al.* 1998. *Anim. Conserv.*, **1**: 129-137.

Barrett, R.H and G.H. Birmingham. 1994. **In**: Prevention and control of wildlife damage. Cooperative Extension Service, University of Nebraska, Lincoln, Nebraska, USA, pp. 65-70.

Barry, P. 1860. The fruit garden. C.M. Saxton, Barker and Co., New York, 398 p.

Bashir, E.A. 1978. *Proc. Seminar on Bird Pest Problems in Agric., FAO/Vertebr. Pest Control Centre, Pakistan Agric. Res. Council,* **1**: 22-27.

Bashir, E.A. 1979. *Proc. Seminar on Bird Pest Problems in Agric., FAO/Vertebr. Pest Control Centre, Pakistan Agric. Res. Council,* **2**: 167-171.

Bateman, J.A. 1971. Animal traps and trapping. Stackpole Books, Harrisburg, Pennsylvania, 286 p.

Baxter, A. 2000. *Proc. International Bird Strike Committee,* IBSC25/WP-AV9, pp. 401-409.

Baxter, A.T and J.R. Allan. 2006. *Wildl. Soc. Bull.*, **34**: 1162-1168.

Bayart, F., K.T. Hayashi, K.F. Faull, J.D. Barchas and S. Levine. 1990. *Behav. Neuro.*, **104**: 98-107.

Beason, R.C. 2004. *Proc. Vertebr. Pest Conf.*, **21**: 92-96.

Beauchamp, G.K and J.R. Mason. 1991. **In**: The hedonics of taste. Lawrence Erlbaum Associates, Hillsdale, New Jersey, pp. 159-184.

Belant, J.L and S.K. Ickes. 1996. *Proc. Vertebr. Pest Conf.*, **17**: 108-112.

Belant, J.L and S.K. Ickes. 1997. *Proc. Great Plains Wildl. Damage Control Workshop*, **13**: 73-80.

Belant, J.L and T.W. Seamans. 1999. *J. Wildlife Diseases*, **35**: 239-242.

Belant, J.L. 1997. *Landscape Urban Plan.*, **38**: 245-258.

Belant, J.L., L.A. Tyson, T.W. Seamans and S.K. Ickes. 1997a. *J. Wildl. Res.*, **2**: 210-212.

Belant, J.L., L.A. Tyson, T.W. Seamans and S.K. Ickes. 1997d. *J. Wildl. Manage.*, **61**: 917-924.

Belant, J.L., S.K. Ickes and T.W. Seamans. 1997b. Federal Aviation Administration Interim Report DTFA01 -91-2-02004, Task 3, Part 11, Experiment 2, Atlantic City, New Jersey, USA.

Belant, J.L., S.K. Ickes, L.A. Tyson and T.W. Seamans. 1997c. *Crop Protec.*, **16**: 439-447.

Belant, J.L., T.W. Seamans and C.P. Dwyer. 1996. *Crop Protec.*, **15**: 575-578.

Belant, J.L., T.W. Seamans and C.P. Dwyer. 1998c. *Int. J. Pest Manage.*, **44**: 247-249

Belant, J.L., T.W. Seamans and L.A. Tyson. 1998a. *Proc. Vertebr. Pest Conf.*, **18**: 107-110.

Belant, J.L., T.W. Seamans and L.A. Tyson. 1998b. *Proc. Vertebr. Pest Conf.*, **18**: 359-362.

Belant, J.L., T.W. Seamans, L.A. Tyson and S.K. Ickes. 1996. *J. Wildl. Manage.*, **60**: 923-928.

Belden, R.C and W.B. Frankenberger. 1990. *Proc. Annual Conf. Southeastern Assoc. Fish Wildl. Agencies*, **44**: 231-249.

Bell, R.H.V. 1984. In: Conservation and wildlife management in Africa. US Peace Corps, Malawi, pp. 387-416.

Belton, P. 1976. Effects of interrupted light on birds. National Research Council, Canada, 73 p.

Benson, A.B. 1966. Peter Kalm's travels in North America. Vol. 1. Dover Publications, Inc., New York, 797 p.

Bentley, W.J and M. Viveros. 1992. *California Agric.*, **46**: 30-32.

Berge, A., M. Delwiche, W.P. Gorenzel and T. Salmon. 2007. *Am. J. Enol. Vitic.*, **58**: 135-143.

Berger, J., P.B. Stacey, L. Bellis and M.P. Johnson. 2001. *Ecol. Appl.*, **11**: 947-960.

Beri, Y.P., M.G. Jotwani, S.S. Misra and D. Chander. 1969. *Indian J. Entomol.*, **30**: 69-71.

Beringer, J., K.C. VerCauteren and J.J. Millspaugh. 2003. *Wildl. Soc. Bull.*, **31**: 492-498.

Beringer, J., L.P. Hansen, R.A. Heinen and N.F. Giessman. 1994. *Wildl. Soc. Bull.*, **22**: 627-632.

Berkhoudt, H. 1985. **In**: Form and function in birds. Vol. 3, pp. 463-496.

Berny, P.J., T. Buronfosse, F. Buronfosse, F. Lamarque, G. Lorgue. 1997. *Chemosphere*, 35: 1817-1829.

Berrill, F.W. 1956. *Queensland Agric. J.*, 82: 435-439.

Berry, C.E. 1938. *Calif. Dept. of Agri. Bull.*, 27: 172-180.

Besser, J.F and J.F. Welch. 1959. *Trans. North Am. Wildl. Conf.*, 24: 166-173.

Besser, J.F. 1976. *Proc. Vertebr. Pest Conf.*, 7: 11-16.

Besser, J.F. 1978. *Proc. Vertebr. Pest Conf.*, 8: 51-55

Besser, J.F. 1985. *U.S. Fish and Wildl. Service Bird Damage Report No.* 340, 90 p.

Bieber, C and T. Ruf. 2005. *J. Appl. Ecol.*, 42: 1203-1213.

Bishop, J., H. McKay, D. Parrott and J. Allan. 2003. Assessed through http:// archive.defra.gov.uk/environment/quality/noise/research/birdscaring/ birdscaring.pdf

Bivings, A.E. 1986. *Proc. Great Plains wildl. Damage Control Workshop*, 7: 64-66.

Bjorge, R.R and J.R. Gunson. 1985. *J. Range Manage.*, 38, 483-487.

Black, H.C., E.J. Dimock, W.E. Dodge and W.H. Lawrence. 1969. *Trans. North Am. Wildl. and Nat. Resour. Conf.*, 34: 388-408.

Black, H.C., O.H. Hewitt and C.W. Severinghaus. 1959. *N.Y. Fish and Game J.*, 6: 179-203.

Black, H.L. 1981. *Rangelands*, 3: 235-237.

Blackshaw, J.K., G.E. Cook, P. Harding, C. Day, W. Bates, J. Rose and D. Bramham. 1990. *Appl. Anim. Behav. Sci.*, 25: 1-8.

Blackwell, B.F. 2002. *Proc. Vertebr. Pest Conf.*, 20: 146–152.

Blackwell, B.F., D.A. Helon, and R.A. Dolbeer. 2001. *Crop Protec.*, 20: 65-68.

Blackwell, B.F., G.E. Bernhardt and R.A. Dolbeer. 2002. *J. Wildl. Manage.*, 66: 250-258.

Blackwell, B.F., T.W. Seamans and R.A. Dolbeer. 1999. *J. Wildl. Manage.*, 63: 1336-1343.

Blokpoel, H. 1976. Bird hazards to aircraft: Problems and prevention of bird-aircraft collisions, Clarke, Irwin and Company Ltd., Toronto, Ontario, Canada, 235 p.

Blumstein, D.T and K.B. Armitage. 1997. *Anim. Behav.*, 53: 143-171.

Boag, D.A and V. Lewin. 1980. *J. Wildl. Manage.*, 44: 145-154.

Bock, M and H. Jackson. 1957. *Br. J. Pharmacol. Chemother.*, 12: 1-7.

Boddicker, M.L. 1982. *Proc. Vertebr. Pest Conf.*, 10: 50-54.

Bogatich, V. 1967. *Proc. Vertebr. Pest Conf.*, 3: 98-99.

Bollengier, R.M., J.L. Guarino and C.P. Stone. 1973. *Proc. Bird Control Sem.*, 6: 216-220.

Bomford, M and P. O'Brien. 1992. *Proc. Vertebr. Pest Conf.*, **15**: 344-347.

Bomford, M and P.H. O'Brien. 1990. *Wildl. Soc. Bull.*, **18**: 411-422.

Bomford, M and R. Sinclair. 2002. *Emu*, **102**: 29-45.

Bomford, M. 1990. Bureau of Rural Resources Bulletin No. 7. Australian Government Publishing Service, Canberra, 50 p.

Boonstra, R and F.H. Rodd. 1982. *J. Mamm.*, **63**: 672-675.

Booth, T.W. 1983. In: Prevention and control of wildlife damage. University of Nebraska, Lincoln, pp.1-5.

Bordeaux, E.S. 1974. Messengers from ancient civilizations. Academy Books, San Diego, California, 43 p.

Boudreau, G.W. 1968. *Proc. Bird Control Sem.*, **4**: 38-44.

Boudreau, G.W. 1972. *Proc. Vertebr. Pest Conf.*, **5**: 121-123.

Boudreau, G.W. 1975. How to win the war with pest birds. Wildlife Technology, Hollister, California, 174 p.

Boulton, W.J and W.J. Freeland. 1991. *Wildl. Res.*, **18**: 63-73.

Bowerman, A.M and J.E. Brooks. 1971. *J. Wildl. Manage.*, **35**: 618-624.

Bowie, M.H and J.G. Ross. 2006. *New Zealand J. Ecol.*, **30**: 219-228.

Bowmaker, J.K and G.R. Martin. 1985. *J. Comp. Physiol. A: Neuroethol. Sens. Neural Behav. Physiol.*, **156**: 71-77.

Boyle, C.M. 1960. *Nature*, **168**: 5-7.

Braun, C.E. 2005. Techniques for wildlife investigation and management, Sixth edition. The Wildlife Society, Bethesda, Maryland, USA.

Bremner, O.E. 1946. *Calif. Dept. Agric. Bull.*, **35**: 151-153.

Bridgman, C.J. 1976. *Proc. Pan-African Congress*, **4**: 383-387.

Bridgman, C.J. 1969. *Ibis*, **111**: 444.

Brockelman, W.Y and N.K. Kobayashi. 1971. *J. Wildl. Manage.*, **35**: 852-855.

Bromenshenk, J. 2004. Can honeybees assist in landmine detection? Division of Biological Sciences. University of Montana, Missoula, USA.

Brooks, J.E and E. Ahmed. 1990. In: A training manual on vertebrate pest management. Agricultural Research Council, Islamabad, Pakistan, pp. 173–179.

Brooks, J.E. 1962. *Proc. Vertebr. Pest Conf.*, **1**: 227-244.

Brothers, N.P., I.E. Eberhard, G.R. Copson and I.J. Skira. 1982. *Australian Wildl. Res.*, **9**: 477-785.

Brough, T. 1965. Field trials with the acoustical scaring apparatus in Britain. Inst. Nat. Rech. Agron., Paris, pp. 279-286.

Brough, T. 1967. Recent developments in bird scaring on airfields. The problems of birds as pests. Institute of Biology Symposium, **17**: 29-38.

Brough, T. 1968. **In:** The problems of birds as pests. Academic Press, London, pp. 29-38.

Brough, T. 1969. *J. Appl. Ecol.*, **6**: 403-410.

Brown, C.R and M.B. Brown. 1995. **In:** Birds of North America Online. Assessed through http://bna.birds.cornell.edu/bna/species/149>

Brown, K.P. 1997. *Bird Conserv. Int.*, **7**: 399-407.

Brown, K.P., N. Alterio and H. Moller. 1998. *Wildl. Res.*, **25**: 419-426.

Brown, R.G.B. 1974. *Can. Wildl. Serv. Rep. Ser.* No. **27**, 57 p.

Brown, T.L., D.J. Decker, S.J. Riley, J.W. Enck, T.B. Lauber, P.D. Curtis and G.F. Matfeld. 2000. *Wildl. Soc. Bull.*, **28**: 797-807.

Brugger, K.E. 1992. *J. Wildl. Manage.*, **56**: 794-799.

Bruggers, R.L. 1980. *Proc. Vertebr. Pest Conf.*, **9**: 21-28.

Bruggers, R.L. 1982. *Crop Protec.*, **1**: 55-65

Bruggers, R.L., J.E. Brooks and R.A. Dolbeer. 1986. *Wildl. Soc. Bull.*, **14**: 161-170.

Bruggers, R.L., J.E. Brooks, R.A. Dolbeer, P.P. Woronecki, R.K. Pandit, T. Tarimo and M. Hoque. 1986. *Wildl. Soc. Bull.*, **14**: 161-170.

Buckle, A.P. 1986. *J. Hygiene,* **96**: 467-473.

Buckle, A.P., F.P. Rowe and A.R. Husin. 1984. *Tropical Pest Manage.,* **30**: 51-58

Bucyanayandi, J.D., J.M. Bergeron and H. Menard. 1990. *J. Chem. Ecol.*, **16**: 2569-2579.

Bullard, R.W and J.O. York. 1985. **In:** Progress in plant breeding. Butterworths, London, England, pp. 193-222.

Bullard, R.W. 1979. *Proc. Bird Control Sem.*, **6**: 229-234.

Bullard, R.W. 1985. **In:** Semio-chemistry flavors and pheromones. Walter deGruyter, New York, USA, pp. 65-94.

Bullard, R.W., T.J. Leiker, J.E. Peterson and S.R. Kilburn. 1978. *J. Agric. Food Chem.*, **26**: 155-158.

Buller, R.M.L., M. Potter and G.D. Wallace. 1986. *Curr. Top. Microbiol.*, **127**: 319-322.

Burbidge, A.A and N.L. McKenzie. 1989. *Biol. Conserv.*, **50**: 143-198.

Burger, J. 1981. *Biol. Conserv.*, **21**: 231-241.

Burton, R.W. 1918. *J. Bombay Nat. Hist. Soc.*, **25**: 491-493.

Busnel, R.G and J. Giban. 1968. **In:** The problems of birds as pests. Symposia of the Institute of Biology No. 17. Academic Press, London, pp. 17-28.

Bykovskii, V and N.V. Kandybin. 1988. **In:** Rodent pest management. CRC Inc, Boca Raton, pp. 377-389.

Caithness, T.A. 1970. *Proc. World Conf. on Bird Hazards to Aircrew*, Kingston, Ont., Sept. 1969. Nat. Res. Counc., Ottawa, Ont., pp. 93-99

Calvi, C., J.F. Besser, J.W. DeGrazio and D.F. Mott. 1977. *Bird Control Sem.*, 7: 255-258.

Calvin, B.A and W.B. Jackson. 1991. *Proc. Int. Cong. Plant Protec.*, 11: 60-64.

Campbell, D.L and J. Evans. 1988. *Proc. Vertebr. Pest Conf.*, 13: 183-187.

Campbell, K.J., G. Harper, D. Algar, C.C. Hanson, B.S. Keitt and S. Robinson. 2011. In: Island invasives: Eradication and management. IUCN, Gland, Switzerland, pp. 37-46.

Campbell, R.W., T.R. Torgersen, S.C. Forrest and L.C. Youngs. 1981. *Pacific Northwest Forest and Range Experiment Station General Tech.Rep.* **125**, 10 p.

Campo, J.L., M.G. Gil and S.G. Davila. 2005. *Appl. Animal Behav. Sci.*, **91**: 75-84.

Cardinell, H.A and D.W. Hayne. 1945. *Michigan Agric. Expt. Stn. Tech. Bull.*, **198**.

Caslick, J.W and D.J. Decker. 1977. *Cornell Univ. Dept. Nat. Resour. Conserv. Circ.*, **5**, 15 p.

Caslick, J.W. 1980. *Proc. Vertebr. Pest Conf.*, 9: 161-162.

Castelli, P.M and S.E. Sleggs. 2000. *Wildl. Soc. Bull.*, **28**: 385-392.

Cavalcanti, S.M.C and F.F. Knowlton. 1998. *Appl. Animal Behav. Sci.*, **61**: 143-158.

Censky, J and M.S. Ficken. 1982. *The Wilson Bull.*, **94**: 590-593.

Chamberlain, P.A., C. Milton and W.A. Wright. 1981. Assessed through http://digitalcommons.unl.edu/gpwdcwp

Chandler, W.H. 1958. Evergreen Orchards. Lea and Febiger, Philadelphia, PA 535 pp.

Chauhan, N.P.S and R. Singh. 1990. *Proc. Vertebr. Pest Conf.*, **14**: 218-220.

Chauhan, N.P.S. 1999. Assessed through www2.wii.gov.in/envis/ghnp_reports/4_18_chauhan.pdf

Chauhan, N.P.S. 2011. Developing management capabilities for wild pigs damage control in agroecosystems in and around protected areas of India. Final report. Wildlife Institute of India, Dehradun.

Chauhan, N.P.S., K.S. Barwal and D. Kumar. 2009. *Acta Silv. Lign. Hung.*, **5**: 189-197

Chen, G., C.R. Ensor and B. Bohner. 1966. *J. Pharmacol. Exp. Ther.*, **152**: 332-339.

Chhangani, A.K and S.M. Mohnot. 2004. *Primate Report*, **69**: 35-47.

Chipman, R.B., T.L. DeVault, D. Slate, K.J. Preusser, M.S. Carrara, J.W. Friers and T.P. Algeo. 2008. *Proc. Vertebr. Pest Conf.*, **23**: 88-93.

Chitty, D and D.A. Kempson. 1949. *Ecol.*, **30**: 356-342.

Choquenot, D and J. Parkes. 2001. *Biol. Conserv.*, **99**: 29-46.

Choquenot, D. 1990. *J. Mamm.*, **71**: 151-155.

Choquenot, D. 1991. *Ecol.*, **72**: 805-13.

Choquenot, D., J. Hone and G. Saunders. 1999. *Wildl. Res.*, **26**: 251-261.

Choquenot, D., J. McIlroy and T. Korn. 1996. Managing vertebrate pests: Feral pigs. Bureau of Resource Sciences, Australian Government Publishing Service, Canberra, Australia.

Choquenot, D., R.J. Kilgour and B.S. Lukins. 1993. *Wildl. Res.*, **20**: 15-22.

Chubb, W.O. 1959. Protection of sunflowers from depredations by birds. Agr. Res., Special Substation, EFS, Dept. of Agr., Portage La Prairie, Manitoba, Canada. Mimeo, 19 p.

Clark, D.E. 1983. Sunset Gardener's Answer Book. Lane Publishing Co. Menlo, Park, CA, 160 p.

Clark, L and C.A. Smeraski. 1990. *J. Exp. Zool.*, **255**: 22–29.

Clark, L and J.L. Belant. 1998. *Appl. Anim. Behav. Sci.*, **57**: 133-144.

Clark, L and P.S. Shah. 1992. **In**: Chemical signals in vertebrates. Plenum Press, New York, pp. 421-427.

Clark, L. 1997. *J. Wildl. Manage.*, **61**: 1352-1358.

Clark, L., P.S. Shah and J.R. Mason. 1991. *J. Exp. Zool.*, **260**: 310-322.

Cleary, E.C and R.A. Dolbeer. 2005. Assessed through http://digitalcommons.unl.edu/icwdm_usdanwrc/133

Clements, H.F. 1980. Sugarcane crop logging and crop control. Univ. Press of Hawaii, Honolulu, 520 p.

Coates, R.W., M.J. Delwiche, W.P. Gorenzel and T.P. Salmon. 2010. *Human–Wildl. Interactions*, **4**: 130-144.

Coblentz, B.E and D.W. Baber. 1987. *J. Appl. Ecol.*, **24**: 403-418.

Cockrum, E.L. 1947. *J. Mamm.*, **28**: 186.

Coffey, M.A and G.H. Johnston. 1997. *Wildl. Soc. Bull.*, **25**: 433-439.

Coleman, J.D. 1993. *New Zealand J. Zool.*, **20**: 341-345.

Colvin, T.R. 1976. *Proc. Annu. Conf. Southeast. Assoc. of Game and Fish Comm.*, **29**: 450-453.

Conklin, J.S., M.J. Delwiche, W.P. Gorenzel and R.W. Coates. 2009. Assessed through http://digitalcommons.unl.edu/hwi/26

Conover, M.R and R.A. Dolbeer. 1989. *Wildl. Soc. Bull.*, **17**: 441-443.

Conover, M.R. 2002. Resolving human-wildlife conflicts: The science of wildlife damage management. Lewis Publishers, Boca Raton, Florida, USA.

Conover, M.R. 1979. *Proc. Bird Control Sem.*, **8**: 16-24.

Conover, M.R. 1982. *Frontiers Plant Sci.*, **35**: 7-8.

Conover, M.R. 1984. *J. Wildl. Manage.*, **48**: 109-116.

Conover, M.R. 1985. *J. Wildl. Manage.*, **49**: 643-645.

Conover, M.R. 1990. *J. Wildl. Manage.*, **54**: 360-365.

Conover, M.R. 1994. *Proc. Vertebr. Pest Conf.*, **16**: 233-234.

Conry, P.J. 1986. *J. Bombay Nat. Hist. Soc.*, **83**: 499-504.

Cooke, A.S. 1980. *Biol. Conserv.*, **18**: 85-88.

Cooke, B.D and F. Fenner. 2002. *Wildl. Res.*, **29**: 689-706.

Cooper, A.D. 1970. *Int. Biodetn. Bull.*, **6**: 105-107.

Cooper, D.M., E.L. Kershner and D.K. Garcelon. 2005. Institute for Wildlife Studies, Arcata, California, USA.

Coppinger, R., L. Coppinger, G. Langeloh, L. Gettler and J. Lorenz. 1988. *Proc. Vertebr. Pest Conf.*, **13**: 209-214.

Courchamp, F and G. Sugihara. 1999. *Ecol. Appl.*, **9**: 112-123.

Courchamp, F., M. Langlais and G. Sugihara. 2000. *J. Anim. Ecol.*, **69**: 154-164.

Cowan, D.P., A.R. Hardy, J.A. Vaughan and W.G. Christie. 1989. **In**: Mammals as pests. Chapman and Hall, London and New York, pp. 178-186.

Cowan, P. 2000. **In**: The brushtail possum. Lincoln, New Zealand, Manaaki Whenua Press, pp. 262-270.

Cowled, B.D., E. Gifford, M. Smith, L. Staples and S.J. Lapidge. 2006. *Wildl. Res.*, **33**: 427-437.

Cowled, B.D., P. Elsworth and S.J. Lapidge. 2008. *Wildl. Res.*, **35**: 651-662.

Crabb, A.C. 1979. *Proc. Bird Control Sem.*, **8**: 25-30.

Crabb, A.C., T.P. Salmon and R.E. Marsh. 1986. *Proc. Vertebr. Pest Conf.*, **12**: 295-302.

Crabtree, D.G. 1962. *Proc. Vertebr. Pest Conf.*, **1**: 327-362.

Crase, F.T and R.S. DeHaven. 1976. *Proc. Vertebr. Pest Conf.*, **7**: 46-50.

Crase, F.T., C.P. Stone, R.W. DeHaven and D.F. Mott. 1976. U.S.D.I., *Fish and Wildl. Serv., Spec. Sci. Rept. Wildl.*, **157**, 18 p.

Craven, S.R and S.E. Hygnstrom. 1994. **In**: Prevention and control of wildlife damage. Univ. Nebraska Coop. Ext. Sew., Lincoln, pp. 25-40.

Crider, E.D and J.C. McDaniel. 1967. *J. Wildl. Manage.*, **31**: 258-264.

Crocker, D.R and K. Reid. 1993. *Wildl. Soc. Bull.*, **21**: 456-460.

Crocker, D.R and S.M. Perry. 1990. *Ibis,* **132**: 300-308.

Crocker, D.R. 1990. *Proc. Vertebr. Pest Conf.*, **14**: 339-342.

Crocker, D.R., C.B. Scanlon, S.M. Perry. 1993. *Appl. Anim. Behav. Sci.*, **38**: 61-66.

Crockford, J.A., F.A. Hayes, J.H. Jenkins and S.D. Feurt. 1957. *J. Wildl. Manage.*, **21**: 213-220.

Crooks, K.R and M.E. Soule. 1999. *Nature*, **400**: 563-566.

Cumming, D.H and B. Jones. 2005. Elephants in southern Africa: management issues and options, Harare: WWF, WWF-SAPRO Occasional Paper No. 11.

Cummings, J.L., C.E. Knittle and J.L. Guarino. 1986. *Proc. Vertebr. Pest Conf.*, **12**: 286-291.

Cummings, J.L., D.L. Otis and J.E. Davis, Jr. 1992. *J. Wildl. Manage.*, **56**: 349-355.

Cummings, J.L., M.L. Avery, *et al.* 2002. *Wildl. Soc. Bull.*, **30**: 816-820.

Cummings, J.L., M.L. Avery, P.A. Pochop, J.E. Davis Jr, D.G. Decker, H.W. Krupa and J.W. Johnson. 1995. *Crop Protec.*, **14**: 257-259.

Cummings, J.L., R.J. Mason, D.L. Otis, J.E. James Jr. and T.J. Ohashi. 1994. *Wildl. Soc. Bull.*, **22**: 633-638.

Cummings, M.W. 1962. *Proc. Vertebr. Pest Conf.*, **1**: 113-125.

Cummins, J.N and R.E. Byers. 1984. *Hort. Sci.*, **19**: 162.

Cummins, J.M and K. Wodzicki. 1980. *New Zealand J. Zool.*, **7**: 427-434.

Curlewis, J.D., A.S. Whilte, S.I. Loudon and A.S. McNeilly. 1986. *J. Endocrinol.*, **110**: 56-66.

Curtis, P and R. Rieckenberg. 2005. *Proc. Wildl. Damage Manage. Conf.*, **11**: 149-158.

Curtis, P., M. Richmond and C. Fitzgerald. 1995. *Proc. East. Wildl. Damage Control Conf.*, **7**: 172-176.

Curtis, P.D and C.H. Petzoldt. 1994. Control of woodchucks in cabbage plantings. *N.Y.S. Vegetable Conf.*, Liverpool, New York, pp. 48-51.

Curtis, P.D., I.A. Merwin, M.P. Pritts and D.V. Peterson. 1994b. *Hort Sci.*, **29**: 1151-1155.

Curtis, P.D., M.J. Fargione and M.E. Richmond. 1994a. *Proc. Vertebr. Pest Conf.*, **16**: 223-227.

Cuthill, I.C., J.C. Partridge, A.T.D. Bennett, S.C. Church, N.S. Hart and S. Hunt. 2000. *Advan. Stud. Behav.*, **29**: 159-214.

Dahl,H. 1984. Assessed through http://oai.dtic.mil/oai/ oai?verb=getRecord&metadataPrefix=html&identifier=ADP004203.

Daly, J.C. 1980. *Aus. Wildl. Res.*, **7**: 421-432.

Damiba, T.E and E.D. Ables. 1993. *Oryx*, **27**: 97-103.

Damron, B.L., G.M. Prine, and R.H. Harms. 1968. *Poult. Sci.*, **47**: 1648-1650.

Dana, R.H. 1962. *Proc. Vertebr. Pest Conf.*, **1**: 126-143.

Daneke, D and D.G. Decker. 1988. *Proc. Vertebr. Pest Conf.*, **13**: 287-292.

Danner, P.A., E. Ackerman and H.W. Frings. 1964. *J. Acoustic Soc. Am.*, **26**: 731.

Darwin, C. 1839. The Voyage of the Beagle. P.F. Collier and Son Corporation, 524 p.

Davids, R.C. 1960. *Farm J.*, **84**: 64-66.

Davies, T.E., S. Wilson, N. Hazarika, J. Chakrabarty, D. Das, D.J. Hodgson and A. Zimmermann. 2011. *Conserv. Lett.*, **4**: 346-354.

Davis, D.E and W.L. Jensen. 1952. *Trans. North Am. Wildl. Conf.*, **17**: 151-158.

Davis, D.E. 1959. *Anat. Rec*, **134**: 549.

Davis, S.F., C.A. Grover, C.A. Erickson, L.A. Miller and J.A. Bowman. 1987. *Percept. Mot. Skills*, **64**: 1215-1222.

Davis, S.F., L.A. Cunningham, T.J. Burke, M.M. Richard, W.M. Langley and J. Theis. 1986. *Bull. Psychos. Soc.*, **24**: 229-232.

Davison, V.E. 1941. *J. Wildl. Manage.*, **5**: 390-394.

DeBoer, W.F and C.P. Ntumi. 2001. *Pachyderm*, **30**: 57-64.

DeCalesta, D.S and J.P. Hayes. 1979. *Pest Control*, **47**: 18-20.

Decker, D.G and M.L. Avery. 1990. *Proc. Vertebr. Pest Conf.*, **14**: 327-331.

Decker, D.G., M.L. Avery and M.O. Way. 1990. *Proc. Vertebr. Pest Conf.*, **14**: 327-331.

DeGarmo, W.R and J. Gill. 1958. West Virginia whitetails. Conservation Commission Bulletin 4, West Virginia Division of Game Management, Charleston, WV, USA.

DeGrazio, J.W. 1964. *Proc. Vertebr. Pest Conf.*, **2**: 43-49.

DeGrazio, J.W., J.F. Besser, T.J. DeCino, J.L. Guarino and E.W. Schafer, Jr. 1972. *J. Wildl. Manage.*, **36**: 1316-1320.

DeHaven, R.W. 1971. *Rice J.*, **74**: 11-12.

DeMoranville, C., H.A Sandler and D. Nolte. 2000. Assessed through http://scholarworks.umass.edu/cranberrybmp/16

Deodikar, G.B., T.A. Seethalakshmi and M.C. Suryanarayana. 1978. *Indian J. Genet.*, **38**: 372-374.

DeWit, H.C.D. 1967. Plants of the world. E.P. Dutton and Co., Inc., New York. Vol. III, 312 p.

Dhindsa, M.S., H.K. Saini and H.S. Toor. 1992. *Trop. Pest Manage.*, **38**: 98-102.

Dickman, C.R. 1996. *Wildl. Biol.*, **2**: 185-195.

Dilks, P., L. Shapiro, T. Greene, M.J. Kavermann, C.T. Eason and E.C. Murphy. 2011. *New Zealand J. Zool.*, **38**: 143-150.

DiSalle, E., G. Ornati, D. Guidici and G. Briatico. 1983. *Acta Endocrinol.*, **103**: 256-265.

Dix, I.D., S.E. Jolly, L.S. Bufton and A.I. Gardiner. 1994. *Wildl. Res.*, **21**: 49-51.

Doggett, H. 1957. *Field Crop Abstr.*, **10**: 153-156.

Doggett, H. 1970. Sorghum. Longmans, Green and Co. Ltd., London, England, 403 p.

Dolbeer, R.A and S.K. Ickes. 1994. *Proc. Vertebr. Pest Conf.*, **16**: 279-282.

Dolbeer, R.A. 1980. *U.S. Fish and Wildl. Service Resource Publ.* 136. Washington, DC, 18 p.

Dolbeer, R.A. 1981. *Wildl. Soc. Bull.,* **9**: 44-51.

Dolbeer, R.A. 1990. *Ibis,* **132**: 309-322.

Dolbeer, R.A., C.R. Ingram and A.R. Stickley Jr. 1973. *Proc. Bird Control Sem.,* **6**: 28-40.

Dolbeer, R.A., J.L. Belant and J. Sillings. 1993. *Wildl. Soc. Bull.,* **21**: 442-450.

Dolbeer, R.A., P.P. Woroneckei and R.L. Bruggers. 1986a. *Wildl. Soc. Bull.,* **14**: 418-425.

Dolbeer, R.A., P.P. Woronecki and J.R. Mason. 1988. *J. Amer. Soc. Hort. Sci.,* **113**: 460-464.

Dolbeer, R.A., P.P. Woronecki and R.A. Stehn. 1982. *Prot. Ecol.,* **4**: 127-139.

Dolbeer, R.A., P.P. Woronecki and R.A. Stehn. 1986b. *J. Amer. Soc. Hort. Sci.,* **111**: 306-311.

Dolbeer, R.A., P.P. Woronecki and R.W. Bullard. 1992. **In**: Chemical signals in vertebrates. Plenum Press, New York, pp. 323-330.

Dolbeer, R.A., R.B. Chipman, A.L. Gosser and S.C. Barras. 2003. *Proc. 26th International Bird Strike Committee meeting.* WPBB5.

Dolbeer, R.A., T.W. Seamans, B.F. Blackwell and J.L. Belant. 1998. *J. Wildl. Manage.,* **62**: 1557-1563.

Dorrance, M.J and L.D. Roy. 1978. *Proc. Vertebr. Pest Conf.,* **8**: 251-254.

Dowson, V.H.W. 1982. *FAO Plant Prodn Protec.,* **35**: 224-227.

Dudderar, G.R., S. Tellman and D.K. Elshoff. 1995. *Eastern Wildl. Damage Manage. Conf.,* **17**: 149-152.

Duffield, J.W and R.P. Eide. 1962. *J. Forest.,* **60**: 109-111.

Duncan, C.J. 1960. *Ann. Appl. Biol.,* **48**: 409-414.

Duncan, G.W and C. Lyster. 1963. *Fertility and Sterility,* **14**: 565.

Dwyer, G., S.A. Levin and L. Buttel. 1990. *Ecol. Monogr.,* **60**: 423-447.

Eadie, W.R. 1961. Bull. No. 1055. New York State Agric. Exp. Sta., Ithaca, 15 p.

Eason, C., A. Miller, S. Ogilvie and A. Fairweather. 2011. *New Zealand J. Ecol.,* **35**: 1-20.

Eason, C.T and M. Wickstrom. 2001. Vertebrate pesticide toxicology manual: Information on poisons used as vertebrate pesticides, 2nd Edition, Department of Conservation, Wellington, New Zealand, 122 p.

Eason, C.T., E.C. Murphy, S. Hix and D.B. MacMorran. 2010b. *Integrative Zool.,* **5**: 31-36.

Eason, C.T., L. Shapiro, P. Adams, S. Hix, C. Cunningham, D. MacMorran, M. Statham and H. Statham. 2010a. *Wildl. Res.,* **37**: 497-503.

Eason, C.T., R. Gooneratne, G.R. Wright, R. Pierce and C.M. Frampton. 1993. *Proc. New Zealand Plant Protec. Conf.*, **46**: 297-301.

Eason, C.T., S.C. Ogilvie, A. Miller, R.J. Henderson, L. Shapiro, S. Hix, D. MacMorran and E. Murphy. 2008. *Proc. Vertebr. Pest Conf.*, **23**: 148-153.

Eastland, W.G and S.L. Beasom. 1986. *Wildl. Soc. Bull.*, **14**, 234-235.

Elder, W.H. 1964. *J. Wildl. Manage.*, **28**: 556-575.

ElHani, A and M.R. Conover. 1995. Assessed through http://digitalcommons.unl.edu/nwrcrepellants/14.

Endepols, S., C.V. Prescott, N. Klemann and A.P. Buckle. 2007. *Int. J. Pest Manage.*, **53**: 285-290.

Engeman, R.M and G.W. Witmer. 2000. *Integr. Pest Manage. Rev.*, **5**: 41-55.

Engeman, R.M., J. Peterla and B. Constantin. 2002. *Int. Biodet. Biodeg.*, **49**: 175-178.

Erickson, A.W. 1957. *Trans. North Am. Wildl. Conf.*, **22**: 520-543.

Erickson, D.W and N.F. Giessman. 1989. *Wildl. Soc. Bull.*, **17**: 544-548.

Erickson, W.A. 1979. *Proc. Bird Control Sem.*, **8**: 185-200.

Erickson, W.A., R.E. Marsh and T.P. Salmon. 1990. *Proc. Vertebr. Pest Conf.*, **14**: 314-316.

Erlinge, S., G. Goransson, G. Hogstedt, G. Jansson, O. Liberg, J. Loman, T. Nilsson, T. VonSchantz and M. Sylven. 1988. *Am. Nat.*, **132**: 148-154.

Erlinge, S., G. Goransson, L. Hansson, *et al.* 1983. *Oikos*, **40**: 36-52.

Erwin, K.L. 1999. Assessed through http://birdstrike.bcrescue.org/klece.html

Evans A.L., V. Sahlen, *et al.* 2012. *Plos One*, **7**: e40520

Evans, R.R., J.W. Gertz and C.T. Williams. 1975. *J. Wildl. Manage.*, **39**: 630-634.

Ezealor, A.U and R.H. Giles Jr. 1997. *Int. J. Pest Manage.*, **43**: 97-104.

Fagerstone, K.A., L.A. Miller, J.D. Eisemann, J.R. O'Hare and J.P. Gionfriddo. 2008. *Wildl. Res.*, **35**: 586-592.

Fagerstone, K.A., P.J. Savarie, D.J. Elias and E.W. Schafer Jr. 1994. *Proc. Science Workshop on 1080*. Miscellaneous Series 28. Wellington, The Royal Society of New Zealand, pp. 33-38.

Fairaizl, S.D. 1992. *Proc. Vertebr. Pest Conf.*, **15**: 105-109.

Faust, B.F., M.H. Smith and W.B. Wray. 1971. *Ann. Zool. Fennici*, **8**: 7.

Feare, C.J. 1989. *Agric. Zool. Rev.*, **3**: 317-342.

Feare, C.J., P.W. Greig-Smith and I.R. Inglis. 1986. In: Acta XIX Congressus Internationalis Ornitholgii, Vol. 1. Univ. Ottawa Press, Ottawa, Cannada, pp. 493-506.

Fellows, D.P and P.W.C. Paton. 1988. *Proc. Vertebr. Pest Conf.*, **13**: 315-318.

Fenner, F and B. Fantini. 1999. Biological control of vertebrate pests: The history of myxomatosis, an experiment in evolution. CABI Publishing, Wallingford, UK, 339 p.

Fenner, F and F.N. Ratcliffe. 1965. Myxomatosis. Cambridge University Press, 379 p.

Fenner, F. 1949. *J. Immunol.*, **63**: 341-373.

Fenner, F. 1983. *Proc. R. Soc. Lond. B*, **218**: 259-285.

Fichtel, C and K. Hammerschmidt. 2002. *Ethology*, **108**: 763-777.

Fick, G.N. 1978. **In:** Sunflower science and technology. Am. Soc. Agron., Madison, Wisconsin, pp. 279-338.

Fisher, P., C. O'Connor and G. Morriss. 2008. *J. Wildl. Diseases*, **44**: 655-663.

Fisher, P., C. O'Connor, G. Wright and C.T. Eason. 2003. Assessed through http://www.doc.govt.nz/documents/science-and-technical/dsis139.pdf

Fisher, P., C.O'Connor, G. Wright and C.T. Eason. 2004. Assessed through http://www.doc.govt.nz/documents/science-and-technical/dsis188.pdf

Fitzwater, W.D. 1972. *Proc. Verteb. Pest Conf.*, **5**: 49-55.

Fitzwater, W.D. 1978. Bird problems ? What you can do about them. Biological consultants, Albuquerque, New Mexico. 61 pp.

Fitzwater, W.D. 1982. *Proc. Vertebr. Pest Conf.*, **10**: 16-20.

Fitzwater, W.D. 1988. *Proc. Vertebr. Pest Conf.*, **13**: 254-259.

Fitzwater, W.D. 1994. The Handbook: Prevention and Control of Wildlife Damage. Paper 71. http://digitalcommons.unl.edu/icwdmhandbook/71

Floyd, J. 1960. Crop Damage by Deer and Bear. Information and Education Division, Florida Game and Fresh Water Fish Commission, Tallahassee, 4 p.

Foran, B.D., W.A. Low and B.W. Strong. 1985. *Aust. Wildl. Res.*, **12**: 237-247.

Forsyth, D.M. 1999. *J. Appl. Ecol.*, **36**: 351-362.

Foster, T.S. 1979. *Proc. Bird Control Sem.*, **8**: 254-255.

Fox, G.J. and G. Linz. 1983. *Proc. Bird Control Sem.*, **9**: 181-189.

Fox, J.R and M.R. Pelton. 1977. **In:** Research and management of wild hog population. Georgetown: Belle Baruch Forest Science Institute, pp. 53-66.

Frank, J.F. 1948. *Can J. Comp. Med.*, **48**: 216-218.

Freeland, W.J and D. Choquenot. 1990. *Ecology*, **71**: 589-597.

Freeland, W.J and D.J. Janzen. 1974. *Am. Nat.*, **108**: 269-289.

Frings, H and J. Jumber. 1954. *Science*, **119**: 318.

Frings, H and M. Frings. 1967. **In:** Pest control: Biological, physical and selected chemical methods. Academic Press, New York, pp. 387-454.

Frings, H. 1964. *Proc. Vertebr. Pest Conf.*, **2**: 50-56.

Frings, H., M. Frings, B. Cox and L. Peissner. 1955. *Science,* **121**: 340-341.

Fukuda, Y., C.M. Frampton and G.J. Hickling. 2008. *New Zealand J. Zool.,* **35**: 217-224.

Fulk, G.W. 1972. *J. Mammal.,* **53**: 461-478.

Fuller-Perrine, L.D and M.E. Tobin. 1993. *Wildl. Soc. Bull.,* **21**: 47-51.

Fungo, B. 2011. *Environ. Res. J.,* **5**: 87-92.

Ganchrow, D and J.R. Ganchrow. 1985. *Physiol. Behav.,* **34**: 889-894.

Gao, Y and R.V. Short. 1993. *J. Reprod. Fert.,* **97**: 39-49.

Garcia, J and W.G. Hankins. 1975. **In**: Olfaction and taste. Academic Press, New York, pp. 39-45.

Garlough, F.E and D.A. Spencer. 1944. *USDI Fish and Wildlife Service Bull.,* **36**: 37.

Garner, K.M. 1978. *Proc. Vertebr. Pest Conf.,* **8**: 54-59.

Garrett, M.G and W.L. Franklin. 1983. *J. Range Manage.,* **36**: 753-755.

Garrett, T. 1999. Alternative traps. Animal Welfare Institute: Washington, DC, USA

Gaston, W., J.B. Armstrong, W. Arjo and S.H. Lee. 2008. Assessed through http://digitalcommons.unl.edu/feralhog/

Geisser, H and H.U. Reyer. 2004. *J. Wildl. Manage.,* **68**: 939-946.

German, A. 1985. *Phytoparasitica,* **13**: 209-213.

Gilbert, B.K and L.D. Roy. 1977. *Proc. Predator Symposium. Univ. of Montana, Missoula,* pp. 93-102.

Gill, E.L., R.W. Watkins, D.P. Cowan, J.D. Bishop and J.E. Gurney. 1998. *Pestic. Sci.,* **52**: 159-164.

Gill, F.B. 1995. Ornithology, WH Freeman Publishing, New York, 766 p.

Gillies, C.A and R.J. Pierce. 1999. *New Zealand J. Ecol.,* **23**: 183-192.

Gilman, R.M. 1978. *Proc. Vertebr. Pest Conf.,* **8**: 63-66.

Gilsdorf, J.M., S.E. Hygnstrom and K.C. Vercauteren. 2002. *Int. Pest Manage. Rev.,* **7**: 29-45.

Gilsdorf, J.M., S.E. Hygnstrom, K.C. VerCauteren, E.E. Blankenship and R.M. Engeman. 2004. *Wildl. Soc. Bull.,* **32**: 524-531.

Glahn, J.F. 2000. *Proc. Vertebr. Pest Conf.,* **19**: 44-48.

Glahn, J.F., J.R. Mason and D.R. Woods. 1989. *Wildl. Soc. Bull.,* **17**: 313-320.

Godwin, C., J.A. Schaefer, B.R. Patterson and B.A. Pond. 2013. *J. Wildl. Manage.,* **77**: 290-296.

Gordon, C.H. 2003. The impact of elephants on the riverine woody vegetation of Samburu National Reserve, Kenya. Report for Save the Elephants.

Gorenzel, W.P and T.P. Salmon. 1982. *Proc. Vertebr. Pest Conf.,* **10**: 179-185.

Gorenzel, W.P and T.P. Salmon. 1993. *Wildl. Soc. Bull.*, **21**: 334-338.

Gorenzel, W.P and T.P. Salmon. 2008. Bird hazing manual. Techniques and strategies for dispersing birds from spill sites. University of California, Agriculture and Natural Resources Publication, Davis. 21638, 102 p.

Gorenzel, W.P., B.F. Blackwell, G.D. Simmons, T.P. Salmon and R.A. Dolbeer. 2002. *Int. J. Pest Manage.*, **48**: 327-331.

Gorenzel, W.P., D.A. Whisson and P.R. Kelly. 2003. Final Report to the California Department of Fish and Game, Office of Spill Prevention and Response. University of California, Davis, 54 p.

Gorman, M and A. Lamb. 1994. *Animal Welf.*, **3**: 3-12.

Gorman, M.L. 1984. *J. Zool.* (London), **202**: 419-423.

Gorman, M.L. 1986. *J. Trop. Ecol.*, **2**: 187-190.

Goyal, S.K and L.S. Rajpurohit. 2000. *UttarPradesh J. Zool.*, **20**: 55-59.

Graham, M.D and T. Ochieng. 2008. *Oryx*, **42**: 76-82.

Grange, M.L. 1986. *Proc. Vertebr. Pest Conf.*, **12**: 215- 225.

Grant, C.C., R. Bengis, D. Balfour and M. Peel. 2007. Controlling the distribution of elephant. Report on Assessment of South African Elephant Management, 51 p.

Greaves, J.H and P. Ayres. 1969. *J. Hygiene*, **67**: 311-315.

Green, J.S and R.A. Woodruff. 1990. *Proc. Vertebr. Pest Conf.*, **14**: 233-236.

Green, J.S. 1989. *Proc. Eastern Wildl. Damage Control Conf.*, **4**: 83-86.

Green, J.S., R.A. Woodruff and T.T. Tueller. 1984. *Wildl. Soc. Bull.*, **12**: 44-50.

Green, J.S., R.A. Woodruff and W.F. Andelt. 1994. *Proc. Vertebr. Pest Conf.*, **16**: 41-44.

Gregory, N.G., L.M. Milne, A.T. Rhodes, K.E. Littin, M. Wickstrom and C.T. Eason. 1998. *New Zealand Veterinary J.*, **46**: 60-64.

Gregory, R.D. 1991. *J. Animal Ecol.*, **60**: 805-821.

Griffiths, R.E. 1987. **In:** Vertebrate pest control and management materials. 5th Volume. ASTM Spec Tech Publ. 974, Philadelphia, pp. 56-63.

Guale, F.G., E.L. Stair, B.W. Johnson, W.C. Edwards and J.C. Haliburton. 1994. *Veterinary and Human Toxicol.*, **36**: 517-519.

Guarino, J.L. 1972. *Proc. Vertebr. Pest Conf.*, **5**: 108-111.

Gunn, W.W.H. 1973. Experimental research on the use of sound to disperse Dunlin sandpipers at Vancouver International Airport. Rep. from LGL Ltd., Edmonton, Alb. for Assoc. Comm. On Bird Hazards to Aircraft, Nat. Res. Council, Ottawa, Ont., 8 p.

Gupta, A., L. Kant, V. Mahajan, S. Saha and H.S. Gupta. 2009. *Curr. Sci.*, **96**: 467-468.

Gurnell, J and J.R. Flowerdew. 1990. Live trapping small Mammals: A practical guide. Publications of the Mammal Society No. 3. The Mammal Society, Reading.

Gurney, J.E., R.W. Watkins, E.L. Gill and D.P. Cowan. 1996. *Appl. Anim. Behav. Sci.*, **49**: 353-363.

Haag-Wackernagel, D. 2000. *Folia Zool.*, **49**: 25-39.

Hafeez, S., T.H. Khan, T.N. Khan, M. Shahbaz and M. Ahmed. 2008. *J. Agri. Soc. Sci.*, **4**: 92-94.

Hahn, E. 1996. Assessed from http://www.int-birdstrike.org/London_Papers/IBSC23%20WP37.pdf

Hale, J.B. 1973. *Wisconsin Conserv. Bull.*, **38**: 16-17.

Hall, L.S and G.C. Richards. 1987. *Proc. Vertebr. Pest Conf.*, **8**: 279-283.

Hall, T.R., W.E. Howard and R.E. Marsh. 1981. *Wildl. Soc. Bull.*, **9**: 296-298.

Hamilton, P.B., J.R. Stevens, P. Holz, B. Boag, B. Cooke and W.C. Gibson. 2005. *Mol. Ecol.*, **14**: 3167-3175.

Hanski, I., L. Hansson and H. Henttonen. 1991. *J. Anim. Ecol.*, **60**: 353-367.

Hanson, L.B., M.S. Mitchell, J.B. Grand, D.B. Jolley, B.D. Sparklin and S.S. Ditchkoff. 2009. *Wildl. Res.*, **36**: 185-191.

Hanson, L.P and J. Beringer. 1997. *Wildl. Soc. Bull.*, **25**: 484-487.

Hansson, L and I. Hoffmeyer. 1973. *Oikos*, **24**: 477-478.

Hansson, L., R. Gref, L. Lundren and O. Theander. 1986. *J. Chem. Ecol.*, **12**: 1569-1577.

Haque, A.K.M.F and D.M. Broom. 1985. *Agric. Ecosyst. Environ.*, **12**: 219-228.

Harada, W. 1977. Proc. 2nd Sunflower Forum. Fargo, N.D.

Hardenberg, J.D.F. 1965. Clearance of birds on airfields. Inst. Nat. Res. Agron., Paris, 346 p.

Harder, J.D and T.J. Peterle. 1974. *J. Wildl. Manage.*, **38**: 183-196.

Harris, H.B., D.G. Cummins and R.E. Burns. 1970. *Agron. J.*, **62**: 633-635.

Harris, H.B. 1969. Bird resistance in grain sorghum. *Proc. 24th Ann. Corn and Sorghum Res. Conf.* American Seed Trade Assoc., Washington, D.C, pp. 113-122.

Harris, R.E and R.A. Davis. 1998. Assessed from www.tc.gc.ca/civilaviation/publications/tp13029/menu.htm

Harthoorn, A.M and J. Bligh. 1965. *Res. Vet. Sci.*, **6**: 290-299.

Harthoorn, A.M. 1962. *Vet. Rec.*, **74**: 410-411.

Harthoorn, A.M. 1963. *Nature*, **198**: 1116.

Hartley, C.W.S. 1977. The oil palm. Longman, NewYork, 806 p.

Hatch, R.C and D.P. Laflamme. 1989. *Veterinary and Human Toxicol.*, **31**: 105-112.

Hawley, J.E., T.M. Gehring, R.N. Schultz, S.T. Rossler and A.P. Wydeven. 2009. *J. Wildl. Manage.*, **73**: 518-525.

Hayes, R., S. Riffell, R. Minnis and B. Holder. 2009. *Southeastern Naturalist*, **8**: 411-426.

Hayne, D.W and H.A. Cardinell. 1949. *Michigan Agric. Exper. Stn. Quart. Bull.*, **32**: 213-219.

Hearn, C.M., G. Shaw, R.V. Short and M.B. Renfree. 1998. *J. Reprod. Fert.*, **113**: 151-157.

Hedges, S and D. Gunaryadi. 2009 *Oryx*, **44**: 139-146.

Heighway, D.G. 1970. *Proc. World Conf. on Bird Hazards*. Nat. Research Council Canada, Ottawa, pp. 187-194.

Heinrich, J.W and S.R. Craven. 1990. *Wildl. Soc. Bull.*, **18**: 405-410.

Henderson, J and E.L. Craig. 1932. Economic mammalogy. Springfield, 397 p.

Henderson, R.J., C.M. Frampton, M.D. Thomas and C.T. Eason. 1994. *Proc. New Zealand Plant Protec. Conf.*, **47**: 112-116.

Henley, M.D and S.R. Henley. 2007. Population dynamics and elephant movements within the Associated Private Nature Reserves (APNR) adjoining the Kruger National Park. Progress Report to the Associated Private Nature Reserves.

Herman, C.H. 1964. *Proc. Bird Control Sem.*, **2**: 112-121.

Hermann, G and W. Kolbe. 1971. *Pflanzenschutz. Nachr. Bayer*, **24**: 279-320.

Hey, D. 1964. *Proc. Vertebr. Pest Conf.*, **2**: 57-70.

Hill, C.M. 2000. *Int. J. Primatol.*, **21**: 299-315.

Hillman-Smith, A.K.K., E. DeMerode, A. Nicholas, B. Buls and A. Ndey. 1995. *Pachyderm*, **19**: 39-48.

Hinds, L.A and C.H. Tyndale-Biscoe. 1994. *Reprod. Fert. Dev.*, **6**: 705-711.

Hoare, R.E. 1992. *Environ. Conserv.*, **19**: 160-164.

Hoare, R.E. 1995. *Pachyderm*, **19**: 54-63.

Hoare, R.E. 2001a. A decision support system for managing human-elephant conflict situations in Africa. IUCN African Elephant Specialist Group Report, 110 p.

Hoare, R.E. 2001b. *Pachyderm*, **30**: 44-49.

Hobbs, J and F.G. Leon. 1987. *Proc. Eastern Wildl. Damage Control Conf.*, **3**: 143-148.

Hochbaum, H.A., S.T. Dillon and J.L. Howard. 1954. *Trans. North Amer. Wildl. Conf.*, **19**: 176-185.

Hockeynos, G.L. 1958. Bird repellent compositions. *Natl. Pest Control Assn. Tech. Release* **859**, 23 p.

Hofman, P.J., L.G. Smith, D.C. Joyce, G.I. Johnson and G.F. Meiburg. 1997. *Postharvest Biol. Tech.*, **12**: 83-91.

Hogarth, A.M. 1929. The rat: A world menace. J. Bale, Sons and Danielsson Ltd., London, United Kingdom, 112 p.

Holcomb, L.C. 1977. *Proc. Bird control Sem.*, **7**: 275-278.

Holevinski, R.A., P.D. Curtis and R.A. Malecki. 2007. *Human-Wildl. Conflicts*, **1**: 257-264.

Holland, D. 1802. The new and complete universal vermin-killer. A. Hogg, London, United Kingdom, 66 p.

Holler, N.R., P.W. Lefebvre, A. Wilson, R.E. Matteson and G.R. Gutnecht. 1983. *Proc. East. Wildl. Damage Control Conf.*, **2**: 146-154.

Hone, J and B. Atkinson. 1983. *Aust. Wildl. Res.*, **10**: 499-505.

Hone, J. 1994. Analysis of vertebrate pest control. Cambridge University Press, New York, 258 p.

Hooke, A.L., L. Allen and L.K.P. Leung. 2006. *Wildl. Res.*, **33**: 181-185.

Horikawa, A.S. Matsuoka and K. Nakamura. 1988. *Bull. Appl. Ornithol.*, **8**: 63-67.

Horn, E.E. 1949. *Trans. North Amer. Wildl. Conf.*, **14**: 577-585.

Hothem, R.L and R.W. Dehaven. 1982. *Proc. Vertebr. Pest Conf.*, **10**: 171-178.

Houk, A., M.J. Delwich, W.P. Gorenzel, T.P. Salmon. 2004. *Proc. Vertebr. Pest Conf.*, **12**: 130-135.

Howald, G., C.J. Donlan, *et al.* 2007. *Conserv. Biol.*, **21**: 1258-1268.

Howard, J. 1992. *World Wastes*, **35**: 32.

Howard, W.E. 1964. *Proc. Vertebr. Pest Conf.*, **2**: 117-126.

Howard, W.E. 1967. *Proc. Vertebr. Pest Conf.*, **3**: 137-157.

Howard, W.E. 1976. *Proc. Vertebr. Pest Conf.*, **7**: 116-120.

Howard, W.E., R.E. Marsh and C.W. Corbet. 1985. *Acta. Zool. Fennica*, **173**: 191-192.

Howard, W.E., R.E. Marsh and R.E. Cole. 1970. *J. Forestry*, **68**: 220-222.

Howell, P.G. 1984. *Acta Zool. Fenn.*, **172**: 111-113.

Htun, P.T and J.E. Brooks. 1979. *Tropical Pest Manage.*, **25**: 246-250.

Humphrys, S and S.J. Lapidge. 2008. *Wildl. Res.*, **35**: 578-585.

Hunt, C.L. 1985. Descriptions of five promising deterrent and repellent products for use with bears. U.S. Fish and Wildlife Service Report 3. Grizzly Bear Recovery Coordinator, Missoula, Montana, USA.

Hunter, F.A. 1974. *Ann. Appl. Biol.*, **76**: 351-353.

Hussain, I. 1990. **In**: Vertebrate pest management. National Agricultural Research Centre, Agricultural Research Council, Islamabad, Pakistan, pp. 187-191.

Hussain, I., S. Ahmad and A.A. Khan. 1992. *Pakistan J. Zool.*, **24**: 247-249.

Huygens, O.C and H. Hayashi. 1999. *Wildl. Soc. Bull.*, **27**: 959-964.

Hwang, S., J. Shieh, T. Kang and C. Fu. 1985. *Proc. Sem. on the control of squirrel damage to forest trees.* Taiwan University, Nantou, Taiwan, pp. 81-85.

Hygnstom, S.E and S.R. Craven. 1988. *Wildl. Soc. Bull.*, **16**: 291-296.

Hygnstrom, S.E. 1990. *Proc. Vertebr. Pest Conf.*, **14**: 20-24.

Hygnstrom, S.E., K.C. Vercauteren and T.R. Schmaderer. 1994. *Proc. Vertebr. Pest Conf.*, **16**: 293-300.

Inglis, I.R and A.J. Isaacson. 1984. *Behav.*, **90**: 224-240.

Inglis, I.R and A.J. Isaacson. 1987. *Crop Protec.*, **6**: 104-108.

Inglis, I.R. 1980. **In:** Bird problems in agriculture. BCPC Publ., Croydon, England, pp. 121-143.

Inglis, I.R., L.W. Huson, M.B. Marshall and P.A. Neville. 1983. *J. Comp. Ethol.*, **62**: 181-208.

Innes, J., B. Warburton, D. Williams, H. Speed and P. Bradfield. 1995. *New Zealand J. Ecol.*, **19**: 5-17.

Innes, J., R. Hay, I. Flux, P. Bradfield, H. Speed and P. Jansen. 1999. *Biol. Conserv.*, **87**: 201-214.

Iossa, G., C.D. Soulsbury and S. Harris. 2007. *Animal Welf.*, **16**: 335-352.

Jackson, T.P and R.J. vanAarde. 2003. *South African J. Sci.*, **99**: 130-136.

Jackson, W.B. 1978. *Proc. Vertebr. Pest Conf.*, **8**: 47-50.

Jackson, W.B., A. Teklehaimanot and B. Ali. 1978. *EPPO Publ., Ser. B*, **84**: 181-189.

Jacob, J., N.A. Herawati, S.A. Davis, G.R. Singleton. 2004. *J. Wildl. Manage.*, **68**: 1130-1137.

Jacobs, W.W and J.N. Labows. 1979. *Physiol. Behav.*, **22**: 173-178.

Jacobs, W.W. 1978. *Physiol. Behav.*, **20**: 579-588.

Jakel, T., Y. Khoprasert, *et al.* 1999. *Int. J. Parasitol.*, **29**: 1321-1330.

Jaksic, F.M and J.L. Yanez. 1983. *Biol. Conserv.*, **26**: 367-374.

Jenkins, D.J. 2003. Guard animals for livestock protection: Existing and potential use in Australia, NSW Agriculture pub., Orange, Australia, 44 p.

Jessup, D.A., W.E. Clark and K.R. Jones. 1964. *J. Zool. An. Med.*, **15**: 8-10.

Jochle, W and M. Jochle. 1993. *J. Reprod. Fert. Suppl.*, **47**: 419-424.

Johansen, C. 1975. Bear Protection for Bees. EM-4005, Cooperative Extension Service, Washington State University, Pullman, 5 p.

Johnson, E and A.J. Tait. 1983. *Mamm. Rev.*, **13**: 167-172.

Johnson, N.C. 1983. **In:** Prevention and control of wildlife damage. Great Plains Agricultural Council Wildlife Resources Committee and Cooperative Extension Service, University of Nebraska, Lincoln, 640 p.

Johnson, R.J and J.F. Glahn. 1998. Assessed through http://digitalcommons.unl.edu/icwdmother/34.

Johnson, R.J. 2000. Assessed through http://digitalcommons.unl.edu/icwdmother/16

Johnson, R.J., A.E. Koehler and O.C. Burnside. 1982. *Proc. Vetebr. Pest Conf.*, **10**: 205-209.

Johnson, R.J., P.H. Cole and W.W. Stroup. 1985. *J. Wildl. Manage.*, **49**: 620-625.

Johnson, W.V. 1964. *Proc. Vertebr. Pest Conf.*, **2**: 90-96.

Jordan, D.M and M.E. Richmond. 1992. *Proc. East. Wildl. Damage Control Conf.*, **5**: 44-47.

Joshi, A., N. Aardesai, R.S. Sharma, A. Mohammed and C.R. Babu. 1998. Handbook of environment, forest and wildlife protection laws in India, WPSI and Natraj Publishers, Dehradun, pp. 21-24.

Jowett, D. 1967. *Plant Breeding Abst.*, **37**: 85.

Joyce, D.C., D.R. Beasley and A.J. Shorter. 1997. *Aust. J. Experimental Agric.*, **37**: 383-389.

Kagoro-Rugunda, G. 2004. *Afr. J. Ecol.*, **42**: 32-41.

Kahila, G. 1991. *Isr. J. Zool.*, **37**: 188.

Kalam, B. 2005. Endangered elephants. Cabtree Publishing Company, New York, 33 p.

Kalla, N.R. 1976. *Bibliography of Reprod.*, **28**: 261-351.

Kamiss, A and A.K. Turkalo. 1999. Elephant crop raiding in the Dzanga-Sangha reserve, Central African Republic. IUCN African Elephant Specialist Group, 7 p.

Karasov, W.H and D.J. Levey. 1990. *Physiol. Zool.*, **63**: 1248-1270.

Kare, M. 1971. Comparative study of taste. **In**: Handbook of sensory physiology, Chemical Senses, Part 2, Taste. New York: Springer-Verlag.

Karidozo, M and F.V. Osborn. 2005. *Pachyderm*, **39**: 26-32.

Kassa, H and W.B. Jackson. 1979. Assessed through http://digitalcommons.unl.edu/icwdmbirdcontrol/9/

Kehat, M., D. Blumberg and S. Greenberg. 1969. *Isr. J. Agric. Res.*, **19**: 121-128.

Keidar, H., S. Moran and J. Wolf. 1975. Playback of distress calls as a means of preventing losses to agriculture by birds. I. Playback experiment with larks (1970-1974). Report for Israeli Ministry of Agric., 24 p.

Kennelly, J.J., J.I. Garrison and B.E. Johns. 1970. *J. Wildl. Manage.*, **34**: 508-513.

Kent, E and S.P. Grossman. 1968. *Physiol. Behav.*, **3**: 361-362.

Kenward, R.E. 1978. *Ann. Appl. Biol.*, **89**: 277-286.

KeshavaBhat, S and A. Sujatha. 1987. *J. Plant Crops*, **15**: 140-142.

Khan, A.A. 2008. *Pakistan J. Zool.*, **40**: 63-64.

Khan, H.A., S. Ahmad, M. Javed and K. Ahmad. 2011. *Int. J. Agric. Biol.*, **13**: 396-400.

Kierulff, M.C., J.R. Stallings and E.L. Sabato. 1991. *Mammalia*, **55**: 633-625.

Killian, G., L. Miller, J. Rhyan, T. Dees, D. Perry, and H. Doten. 2003. *Proc. Wildl. Damage Manage. Conf.*, **10**: 128-133.

Kimber, D.S. 1963. *Natl. Assoc. Adv. Sci. Q. Rev.*, **15**: 40-41.

King, C. 1989. The natural history of weasels and stoats. Christopher Helm, London, U.K., 253 p

King, F.A and P.C. Lee. 1987. *Primate Conserv.*, **8**: 82-84.

King, L.E, J. Soltis, I. Douglas-Hamilton, A. Savage and F. Vollrath. 2010. *PLoS One*, **5**: e10346.

King, L.E., A. Lawrence, I. Douglas-Hamilton and F. Vollrath. 2009. *Afr. J. Ecol.*, **47**: 131-137.

King, L.E., I. Douglas-Hamilton and F. Vollrath. 2007. *Curr. Biol.*, **17**: 832-833.

King, L.E., I. Douglas-Hamilton and F. Vollrath. 2011. *Afr. J. Ecol.*, **49**: 431-439.

King, N.J., R.W. Mungomery and C.G. Hughes. 1965. Manual of cane-growing. American Elsevier Publishing, New York, 375 p.

Kinnear, J.E., M.L. Onus and R.N. Bromilow, 1988. *Aust. Wildl. Res.*, **15**: 435-450.

Kinnear, J.E., M.L. Onus and N.R. Sumner. 1998. *Wildl. Res.*, **25**: 81-88.

Kinsey, C. 1976. *Minn. Wildl. Res. Quar.*, **36**: 122-138.

Kioko, J., P. Muruthi, P. Omondi and P.I. Chiyo. 2008. *South Afr. J. Wildl. Res.*, **38**: 52-58.

Kitagawa, H., K. Manabe and E.B. Esguerra. 1992: *Acta Hortic.*, **321**: 871-875.

Klemola, T., M. Koivula, E. Korpimaki, *et al.* 1997. *J. Anim. Ecol.*, **66**: 607-614.

Knight, T.A and F.N. Robinson. 1978. *Emu*, **78**: 235-236.

Knittle, C.E and R.D. Parker. 1988. Assessed through https://cowebapps.aphis. usda.gov/wildlife_damage/nwrc/publications/88pubs/knittle881a.pdf

Knittle, C.E and R.D. Porter. 1988. Waterfowl damage and control methods in ripening grain: An overview. U.S. Fish and Wildlife Service tech. Report **14**. Washington, DC, 17 p.

Knowlton, F.F., E.M. Gese and M.M. Jaeger. 1999. *J. Range Mange.*, **52**: 398-412.

Koehler, A.E., R.E. Marsh and T.P. Salmon. 1990. *Proc. Vertebr. Pest Conf.*, **14**: 168-173.

Koehler, J.W. 1962. *Proc. Vertebr. Pest Conf.*, **1**: 174-185.

Koelle, G.B. 1975. **In:** The pharmacological basis of therapeutics. 5th ed., Chap. 28, Macmillan, New York, pp. 575-588.

Kopp, D.D., R.B. Carlson and J.F. Cassel. 1980. Blackbird damage control. North Dakota state University Coop. Ext. Service and U.S. Dept. Agriculture Circular E-692, 5 p.

Korpimaki, E and K. Norrdahl. 1998. *Ecology,* **79**: 2448-2455.

Korpimaki, E., K. Norrdahl, *et al.* 2002. *Proc. R. Soc. Lond. B.,* **269**: 991-997.

Koski, W.R., S.D. Kevan, W.J. Richardson and K. Strom. 1993. Bird Dispersal and Deterrent Techniques for Oil Spills in the Beaufort Sea, Delta Environmental Management Group Ltd., Loveland, Colorodo, 100 p.

Krane, R.V and A.R. Wagner. 1975. *J. Comp. Physiol. Psycol.,* 88: 882-889.

Krzysik, A.J. 1987. A review of bird pests and their management. Technical Report REMREM- 1, 145 p.

Kulkarni, J., M. Prachi and H. Umesh. 2008. Man-elephant conflict in Sindhudurg and Kolhapur districts of Maharashtra, India: Case study of a state coming to terms with presence of wild elephants. Final Technical Report. Envirosearch, Pune.

Kumar, L.S.S., A.C. Aggarwal, H.R. Arakeri, M.G. Kamath, E.N. Moore and R.L. Donahue. 1963. Agriculture in India. Asia Publ. House, Bombay. Vol. II, 243 p.

Lagler, K.R. 1939. *J. Wildl. Manage.,* 3: 169-179.

Langley, W.M., J. Theis, S.F. Davis, M.M. Richard and C.M. Grover. 1987. *Bull. Psychon. Soc.,* **25**: 17-19.

Larkin, R.P., J.R. Torre-Bueno, D.R. Griffin and C. Walcott. 1975. *Proc. Natl. Acad. Sci. USA,* **72**: 1994-1996.

LaVinka, P.C., A. Brand, V.J. Landau, D. Wirtshafter and T.J. Park. 2009. *J. Comp. Physiol. A: Neuroethol. Sens. Neural Behav. Physiol.,* **195**: 419-427.

Lavoie, G.K and J.F. Glahn. 1977. *J. Stored Prod. Res.,* **13**: 23-28.

Lawrence J.H., A.B. Bauer, *et al.* 1975. Bird strike alleviation techniques- Technical discussion. AFFDL-TE-75 -2, volume 1.

Laycock, G. 1966. The alien animals. Natural History Press, Garden City, New York, 240 p.

Lefebrve, P.W and D.E. Mott. 1987. Reducing bird-aircraft hazards at airports through control of bird nesting, roosting, perching, and feeding. Denver Wildlife Research Center, Bird Damage Research Report 390, Denver, Colorado, USA, 90 p.

Lefebrve, L.W. 1978. *Wildl. Soc. Bull.,* 6: 231-234.

Lenghaus, C., H. Westberry, B. Collins, M. Ratnamohan and C. Morrissy. 1994. **In:** Rabbit haemorrhagic disease: Issues in assessment for biological control. Australian Govt Publishing Centre, Canberra, pp. 104-129.

Levett, P.N., D. Walton, L.D. Waterman, C.U. Whittington, G.E. Mathison and C.O.R. Edwards. 1998. *West Indian Medical J.,* **47**: 15-17.

Lewis, J.H. 1946. *J. Wildl. Manage.,* **10**: 277.

Lin, C. 2003. *Micronesica,* 7: 101-112.

Lindell, L.E and A. Forsman. 1996. *Can. J. Zool.,* **74**: 1000-1007.

Lindroth, R.L. 1988. **In**: Chemical mediation of co-evolution. Academic Press, San Diego, pp. 415-445.

Lindsay, S.W., J.K. Levy, S.A. Robertson, A.M. Cistola and L.A. Centonze. 2002. *J. Am. Vet. Med. Assoc.*, **220**: 1491-1495.

Linhart, S.B and R.K. Enders. 1964. *J. Wildl. Manage.*, **28**: 358-363.

Linhart, S.B., G.J. Dasch and F.J. Turkowski. 1981. *Proc. Worldwide Furbearer Conf.*, **8**: 1560-1578.

Linhart, S.B., J.D. Roberts, S.A. Shumake and R. Johnson. 1976. *Proc. Vertebr. Pest Conf.*, **7**: 302-306.

Linhart, S.B., R.T. Sterner, T.C. Carrigan and D.R. Henne. 1979. *J. Range Manage.*, **32**: 238-241.

Linnell, J.D.C., M.E. Smith, J. Odden, P. Kaczensky and J.E. Swenson. 1996. Carnivores and sheep farming and Norway. NINA Opdragsmelding **443**, 118 p.

Linz, G.M., H.J. Homan, A.A. Slowik and L.B. Penry. 2006. *Crop Protec.*, **25**: 842-847.

Lipcius, R.N., C.A. Coyne, B.A. Fairbanks, D.H. Hammond, P.J. Mohan, D.J. Nixon, J.J. Staskiewicz and F.H. Heppner. 1980. *J. Wildl. Manage.*, **44**: 511-518.

Littauer, G. 1990. Frightening techniques for reducing bird damage at aquaculture facilities. Southern Regional Aquaculture Center, Publ. No. **401**, 4 p.

Littauer, G.A. 1993. **In**: Feral swine: A compendium for resource managers. Texas Agricultural Extension Service, San Angelo, Texas, USA, 139-148.

Littin, K.E., N.G. Gregory, A.T. Airey, C.T. Eason and D.J. Mellor. 2010. *Wildl. Res.*, **36**: 709-720.

Liu, S.J., H.P. Xue, B.Q. Pu and N.H. Quain. 1984. *Animal Husbandry and Veterinary Medicine*, **16**: 253-255.

Locombe, D., A. Cyr and J.M. Bergeron. 1986. *J. Appl. Ecol.*, **23**: 773-779.

Logan, K.A., L.L. Sweanor, J.F. Smith and M.G. Hornocker. 1999. *Wildl. Soc. Bull.*, **27**: 201-208.

Logsdon, G. 1981. Organic Orcharding. Rodale Press, Emmaus, PA., 415 p.

Longhurst, W.M., M.B. Jones, R.R. Parks, L.W. Neubaur and M.W. Cummings. 1962. Calif. Agric. Exp. Sta. Ext. Serv. Bull. No. **514**, 17 p.

Lord, W.G. 1979. *American Bee J.*, **119**: 818, 820-821.

Lorenz, J.R. 1985. *Oregon State Univ. Ext. Serv., Ext. Circ.* 1224.

Lorenz, J.R., R.P. Coppinger and M.R. Sutherland. 1986. *J. Range Manage.*, **39**: 293-295.

Lostetter, C.H. 1960. *Trans. North Amer. Wildl. Nat. Resources Conf.*, **25**: 102-109.

Loudon, A.S.I and B.R. Brinklow. 1990. *J. Reprod. Fert.*, **90**: 611-618.

Lu, K.H., L.H. Lee, *et al.* 1994. *Plant Protec. Bull.*, **36**: 209-223.

Lund, M. 1975. **In**: Biological control of rodents. Swedish Natural Science Research Council, Stockholm, pp. 129-137.

Lund, M. 1984. *Pest Control*, **52**: 16.

Lustick, S.I. 1973. *Proc. Bird Control Sem.*, **6**: 171-186.

Lyman, J.H. 1844. *The Amer. Agriculturists*, **3**: 241-242.

Mabbayad, B.B and K.W. Tipton. 1975. *Phil. Agric.*, **59**: 1-6.

Madhusudan, M.D. 2003. *Environ. Manage.*, **31**: 466-475.

Maly, M.S and J.A. Cranford. 1985. *Acta Theriologica*, **30**: 161-165.

Manikowski, S and F. Billiet. 1984. *Trop. Pest Manage.*, **30**: 148-150.

Manser. M.B. 2001. *Proc. R. Soc. Lond. B.*, **268**: 2315-2324.

Mapston, M.E. 1999. *Proc. Feral Swine Symposium*, Fort Worth, Texas, USA, pp. 117-120.

Marks, C.A. 2009. *Wildl. Res.*, **36**: 342-352.

Marks, C.A., F. Gigliotti, F. Busana, M. Johnston and M. Lindeman. 2004. *Anim. Welf.*, **13**: 401-407.

Marks, C.A., M. Nijk, F. Gigliotti, F. Busana and R.V. Short. 1996. *Wildl. Res.*, **23**: 161-168.

Marples, N.M and T.J. Roper. 1997. *Anim. Behav.*, **53**: 1263-1270.

Marsh, A.E., A.E. Koehler and T.P. Salmon. 1990. *Proc. Vertebr. Pest Conf.*, **14**: 174-180.

Marsh, R.E and T.P. Salmon. 1979. *Proc. World Lagomorph Conf.*, pp. 842-857.

Marsh, R.E. 1962. *Proc. Vertebr. Pest Conf.*, **1**: 98-112.

Marsh, R.E. 1977. *EPPO Bull.*, **7**: 495-502.

Marsh, R.E. 1991. *Proc. Great Plains Wildl. Damage Conf.*, **10**: 122-133.

Marsh, R.E., W.A. Erickson and T.P. Salmon. 1991. Bird hazing and frightening methods and techniques. Wildlife Damage Management, Internet Center for Other Publications in Wildlife Management. University of Nebraska – Lincoln, pp. 45-51.

Marsh, R.E., W.A. Erickson and T.P. Salmon. 1992. *Proc. Vertebr. Pest Conf.*, **15**: 112-114.

Martin, C.M and L.R. Martin. 1982. *Proc. Vertebr. Pest Conf.*, **10**: 190-192.

Martin, G. 1986. *Ibis*, **128**: 266-277.

Martin, G.R., W.E. Kirkpatrick, D.R. King, I.D. Robertson, P.J. Hood and J.R. Sutherland. 1994. *Wildl. Res.*, **21**: 85-93.

Martin, L.R. 1976. *Proc. Bird Control Sem.*, **7**: 259-266.

Mason, G and K.E. Littin. 2003. *Anim. Welf.*, **12**: 1-37.

Mason, J.R and D.N. Matthew. 1996. *Int. J. Pest Manage.*, **42**: 47-49.

Mason, J.R and G. Linz. 1997. *Crop Protec.*, **16**: 107-108.

Mason, J.R and J.A. Maruniak. 1983. *Pharmaeol. Biochem. Behav.*, **19**: 857-862.

Mason, J.R and L. Clark. 1990. Denver Wildlife Research Center Bird Section Research Report, **457**, 10 p.

Mason, J.R and L. Clark. 1994. *Crop Protec.*, **13**: 531-534.

Mason, J.R and L. Clark. 1996. *Crop Protec.*, **15**: 97-100.

Mason, J.R and L. Clark. 2000. **In:** Sturkie's avian physiology. 5ᵗʰ ed., Marcel Dekker, Inc., New York, pp. 39-56.

Mason, J.R and R.F. Reidinger. 1983. **In:** Vertebrate pest control and management materials. Philadelphia, Pennsylvania, 315 p.

Mason, J.R and T. Primus. 1996. *Crop Protec.*, **15**: 723-726.

Mason, J.R. 1988. *Proc. Vertebr. Pest Conf.*, **18**: 325-329.

Mason, J.R. 1990. *J. Wildl. Manage.*, **54**: 130-135.

Mason, J.R. 1995. *Int. J. Pest Manage.*, **41**: 19-21.

Mason, J.R. 1998. *Proc. Vertebr. Pest Conf.*, **18**: 325-329.

Mason, J.R., L. Clark and N.J. Bean. 1993. *Crop Protec.*, **12**: 497-500.

Mason, J.R., M.A. Adams and L. Clark. 1989. *J. Wildl. Manage.*, **53**: 55-64.

Mason, J.R., M.A. Adams, R.A. Dolbeer, R.A. Stehn, P.P. Woronecki and G.J. Fox. 1986. *North Dakota Farm Res.*, **43**: 16-20.

Mason, J.R., N.J. Bean, P.S. Shah and L. Clark. 1991. *J. Chem. Ecol.*, **17**: 2539-2552.

Mason, J.R., R.A. Dolbeer, A.H. Arzt, R.F. Reidinger and P.P. Woronecki. 1984. *J. Wildl. Manage.*, **48**: 611-616.

Mason, R.J and L. Clark. 1995. Assessed through http://digitalcommons.unl.edu/nwrcrepellants/26

Mastrota, F.N and J.A. Mench. 1995. *Environ. Toxicol. Chemist.*, **14**: 631-638.

Mattina, M.J.I., J.J. Pignatello and R.K. Swihart. 1991. *J. Chem. Ecol.*, **17**: 451-462.

Mattingly, A. 1976. *Airport Forum*, **4**: 13-28.

May, R.M and R.M. Anderson. 1983. *Proc. R. Soc. Lond. B*, **219**: 281-313.

McAninch, J.B., R.J. Winchcombe and M.R. Ellingwood. 1983. *Proc. East. Wildl. Damage Control Conf.*, **1**: 29-30.

McAtee, W.L. 1939. *J. Wildl. Manage.*, **3**: 1-4.

McCabe, R.A. 1966. **In:** Scientific aspects of pest control, Natl. Acad. Sci. Res. Council Publ. No. **1402**: 115-134.

McCann, B.E and D.K. Garcelon. 2008. *J. Wildl. Manage.*, **72**: 1287-1295.

McComb, K., C. Moss., S. Durant, L. Baker and S. Sayialel. 2001. *Science*, **292**: 491-494.

McCracken, H.F. 1972. *Proc. Vertebr. Pest Conf.*, **5**: 124-126.

McDonald, R.A and S. Harris. 2000. *Mamm. Rev.*, **31**: 57-64

MacDonald, D.W., V.E. Sidorovich, *et al.* 2002. *Ecography*, **25**: 295-302.

McGlone, J.J and J.L. Morrow. 1987. *Appl. Anim. Behav. Sci.*, **17**: 77-82.

McIlroy, J.C and D.R. King. 1990. *Aust. Wildl. Res.*, **17**: 11-13.

McIlroy, J.C and E.J. Gifford. 1997. *Wildl. Res.*, **24**: 483-491.

McIlroy, J.C and R.J. Saillard. 1989. *Aust. Wildl. Res.*, **16**: 353–363.

McIlroy, J.C. 1983. *Aust. Wildl. Res.*, **10**: 139-148.

McKay, H.M., R. Furness, *et al.* 1999. The assessment on the effectiveness of the management measures to control damage by fish eating birds to island fisheries in England and Wales. Report to the Ministry of Agric., Fisheries and Food.

McKay, H.V and D. Parrott. 2002. *Int. J. Pest Manage.*, **48**: 189-194.

McKillop, I.G and C.J. Wilson. 1987. *Wildl. Soc. Bull.*, **15**: 394-401.

McKillop, I.G and R.M. Silby. 1988. *Mamm. Rev.*, **18**: 91-103.

McKillop, I.G., H.W. Pepper and C.J. Wilson. 1986. *Proc. Vetebr. Pest Conf.*, **12**: 147-152.

McKillop, I.G., H.W. Pepper and C.J. Wilson. 1988. *Forestry*, **61**: 359-368.

McKillop, I.G., P. Butt, J. Lill, H.W. Pepper, and C.J. Wilson. 1998. *Crop Protec.*, **17**: 393-400.

MacKintosh, G.R. 1950. *New Zealand J. Agric*, **3**: 4.

McKnight, T.L. 1969. *Geogr. Rev.*, **59**: 330-347.

McLennan, J.A., N.P.E. Langham and R.E.R. Porter. 1995. *New Zealand J. Crop and Horti. Sci.*, **23**: 139-144.

Meadows, L.E and F.F. Knowlton. 2000. *Wildl. Soc. Bull.*, **28**: 614-622.

Meagher, M. 1989. *Wildl. Soc. Bull.*, **17**: 15-19.

Mealey, R.H. 1969. Nylon fencing to protect forest plantations from deer and elk. Wildlife and reforestation in the Pacific Northwest. Oregon State University Corvallis, Oregon, pp. 89-9O.

Meanly, B. 1971. *U.S. Fish and Wildl. Service Resource Publ.* 100. Washington DC, 64 p.

Meehan, A.P. 1976. *Int. Pest Control*, **18**: 12-15.

Meehan, A.P. 1978. *Proc. Vertebr. Pest Conf.*, **8**: 122-126.

Meehan, A.P. 1984. Rats and mice: Their biology and control. The Rentokil Library: East Grinstead, UK.

Melchiors, M.A and C.A. Leslie. 1985. *J. Wildl. Manage.*, **49**: 358-362.

Messersmith, D.H. 1975. *Atl. Nat.*, **30**: 110-114.

Meyer, S. 1991. Being kind to animal pests: A no-nonsense guide to humane animal control with cage traps. Meyer: Garrison, Iowa, USA, 132 p.

Meylan, A. 1978. *Proc. Vertebr. Pest Conf.*, **8**: 73-77.

Mikx, F.H.M. 1970. *Proc. World Conf. on Bird Hazards*, Nat. Research Council Canada, Ottawa, pp. 203-205.

Millar, S.E., S.M. Chamow, A.W. Baur, C. Oliver, F. Robey and J. Dean. 1989. *Science*, **246**: 935-938.

Minta, S.C and R.E. Marsh. 1988. *Proc. Vertebr. Pest Conf.*, **13**: 199-208

Mitchell, R.T. and J.T. Linehan. 1967. *USDI, Fish and Wildl. Serv.,Wildl.*, **476**: 8.

Moen, A.N. 1983. Agriculture and wildlife management. Corner Brook Press, Lansing, New York, 367 p.

Montevecchi, W.A and A.D. Maccarone. 1987. *J. Field Ornithol.*, **58**: 148-151.

Monticelli, P.F., R.S. Tokumaru, C. Ades. 2004. *Ann. Braz. Acad. Sci.*, **76**: 368-372.

Montoney, A.J and H.C. Boggs. 1995. *Eastern Wildl. Damage Control Conf.*, **6**: 59-66.

Moore, C.A and R. Elliott. 1946. *J. Comp. Neurol.*, **84**: 119-131.

Moran, S. 1991. *Phytoparasitica*, **19**: 95-101.

Morgan, D.R., L. Milne, C. O'Connor and W.A. Ruscoe. 2001. *Int. J. Pest Manage.*, **47**: 277-284.

Morris, P and S. Whitbread. 1986. *J. Zool.* (London), **210**: 642-644.

Morris, R.D. 1968. *Canadian Field-Naturalist*, **82**: 84-87.

Moseby, K.E and J.L. Read. 2006. *Biol. Conserv.*, **127**: 429-437.

Moseby, K.E and B.M. Hill. 2011. *Wildl. Res.*, **38**: 338-349.

Mossler, K. 1980. *Proc. Bird Strike Commit Europe.* The Hague **14**, 58 p.

Mott, D.F and S.K. Timbrook. 1988. *Proc. Vertebr. Pest Conf.*, **13**: 301-304.

Mott, D.F. 1980. *Proc. Vertebr. Pest Conf.*, **9**: 38-42.

Mott, D.F. 1985. *Proc. Eastern Wildl. Damage Control Conf.*, **2**: 156-162.

Mott, D.F., W.C. Royall Jr., C.F. Knittle, O.E. Bray and J.L. Guarino. 1977. Denver Wildl. Res. Center, Bird Damage Res. Rep. No. **31**, 11 p.

Mulla, M.S and T. Su. 1999. *J. American Mosquito Control Assoc.*, **15**: 133-152

Muller-Schwarze, D. 1972. *J. Mamm.*, **53**: 393-394.

Munoz, A and R. Murua. 1990. *Acta Oecologica*, **11**: 251-261.

Murphy, E.C., B.K. Clapperton, P.M.F. Bradfield and H.J. Speed. 1998. *New Zealand J. Zool.*, **25**: 315-328.

Murphy, E.C., L. Robbins, J.B. Young and J.E. Dowding. 1999. *New Zealand J. Ecol.*, **23**: 175-182.

Murphy, E.C., L. Shapiro, S. Hix, D. MacMorran and C.T. Eason. 2011. **In**: Island invasives: Eradication and management. IUCN, Gland, Switzerland, pp. 213-216.

Murton, R.K., N.J. Westwood and A.J. Isaacson. 1974. *J. Appl. Ecol.*, **11**: 61-81.

Murton, R.K., R.J.P. Thearle and J. Thompson. 1972. *J. Appl. Ecol.*, **9**: 835-74.

Murua, R and J. Rodriguez. 1989. *J. Appl. Ecol.*, **26**: 81-88.

Myers, A.K. 1967. *J. Comp. Physiol. Psychol.*, **64**: 171-175.

Myers, K., I.D. Marshall and F. Fenner. 1954. *J. Hygiene*, **52**: 337-362.

Naef-Daenzer, L. 1983. *Proc. Bird Control Sem.*, **9**: 91-95.

Nakamura, K. 1997. *J. Wildl. Manage.*, **61**: 925-934.

Nakamura, K., Y. Shizota, T. Kaneko and S. Matsuoka. 1995. *Appl. Entomol. Zool.*, **30**: 383-392.

Neal, B.R and A.G. Cock. 1969. *J. Zool.* (London), **158**: 335-340.

Neal, E. 1986. Badgers. Croom Helm, London, 238 p.

Neff, J.A and B. Meanley. 1957. *Univ. Arkansas Agric. Exp. Sta. Bull.*, **584**: 89.

Neff, J.A. 1948. Protecting crops from damage by horned larks in California. *U.S. Fish and wildl. service*, Washington, DC, 11 p.

Neff, J.A. 1979. Frightening blackbirds from rice fields. U.S. Fish and wildl. Service and University of Arkansas College of Agric., 7 p.

Nelson, A., P. Bidwell and C. Sillero-Zubiri. 2003. A review of humane elephant conflict management strategies. People and Wildlife Initiative. Wildlife Conservation Research Unit, Oxford University.

Nelson, A., P. Bidwell and C. Sillero-Zubiri. 2007. A review of human-elephant conflict management strategies. People and Wildlife Initiative. Wildlife Conservation Research Unit, Oxford University.

Nelson, P.C and G.J. Hickling. 1994. *Proc. Vertebr. Pest Conf.*, **16**: 217-222.

Nelson, S.O and J.L. Seubert. 1966. **In**: Pest control by chemical, biological, genetic and physical means. ARS, USDA, pp. 177-194.

Nelson, T.S., E.L. Stephenson, A. Burgos, J. Floyd and J.O. York. 1975. *Poult. Sci.*, **54**: 1620-1623.

Nettles, V.F., J.L. Corn, G.A. Erickson and D.A. Jessup. 1989. *J. Wildl. Dis.*, **25**: 61-65.

Newmark, W.D., N.L. Leonard, H.I. Sariko and D.M. Gamasa. 1993. *Biol. Conserv.*, **63**: 177-183.

Newsome, A.A. 1990. *Trends Ecol. Evol.*, **5**: 187-191

Newsome, A.E., R.P. Pech, R. Smyth, P. Banks and C. Dickman. 1997. Potential impacts on the Australian native fauna of rabbit calicivirus disease. Canberra: Biodiversity group, Environment, Australia, Canberra, 130 p.

Newton, I., I. Wylie and P. Freestone. 1990. *Environ. Pollution*, **68**: 101-117

Ngure, N. 1995. *Pachyderm*, **19**: 20-25.

Nichols, E.L and S.L. Balloun. 1962. *Poultry Sci.*, **41**: 1982-1984.

Nolte, D.L., J.P. Farley, D.L. Campbell, G. Epple and J.R. Mason. 1993a. *Pestic. Sci.*, **12**: 624-626.

Nolte, D.L., J.R. Mason and L. Clark. 1993b. *J. Chem. Ecol.*, **19**: 2019-2017.

Nolte, D.L., J.R. Mason and S.L. Lewis. 1994b. *J. Chem. Ecol.*, **20**: 303-308.

Nolte, D.L., J.R. Mason, G. Epple, E. Aronov and L. Campbell. 1994a. *J. Chem. Ecol.*, **20**: 1505-1516.

Nomsen, D.E. 1989. **In**: Waterfowl, ripening grain damage and control methods. U.S. Fish Wildl. Serv., Washington, DC.

Norbury, G. 2000. *New Zealand J. Zool.*, **27**: 145-163.

Norrdahl, K and E. Korpimaki. 1995. *Proc. R. Soc. Lond. B.*, **261**: 49-53.

Nugent, G and K.W. Fraser. 1993. *New Zealand J. Zool.*, **20**: 361-366.

Nyhus, P.J., R. Tilson and Sumianto. 2000. *Oryx*, **34**: 262-274.

Ochse, J.J. 1931. Fruits and fruitculture in the Dutch East Indies. G. Kolff and Co., Batavia-C, 180 p.

O'Connell, R.A and J.P. Clark. 1992. *Proc. Vetebr. Pest Cont.*, **15**: 326-329.

Oguge, N., D. Ndungu and P. Okemo. 1997. *Belgian J. Zool.*, **127**: 129-135.

Oh, H.K., T. Sakai, M.B. Jones and W.M. Longhurst. 1967. *Appl. Microbiol.*, **15**: 777-784.

Ohgushi, R. 1986. *Appl. Entomol. Zool.*, **21**: 627-629.

Oliver, A.J and D.R. King. 1983. *Aust. Wildl. Res.*, **10**: 297-301.

Oliver, A.J., S.H. Wheeler and C.D. Gooding. 1982. *Aust. Wildl. Res.*, **9**: 125-134.

Omondi, P., E. Bitok and J. Kagiri. 2004. *Pachyderm*, **36**: 80-86.

Onclin, K., L.D.M. Silva, I. Donnay and J.P. Verstegen. 1993. *J. Reprod. Fert. Suppl.*, **47**: 403-409.

Orr-Walker, T., N.J. Adams, L.G. Roberts, J.R. Kemp and E.B. Spurr. 2012. *Appl. Animal Behav. Sci.*, **137**: 80-85

Osborn, F.V and G. Parker. 2003. *Orynx*, **37**: 80-84.

Osborn, F.V and L.E.L. Rasmussen. 1995. *Pachyderm*, **20**: 55-64.

Osborn, F.V and S. Anstey. 2002. Elephant/human conflict and community development around the Niassa Reserve, Mozambique. Mid Zambezi Elephant project, Consultancy for WWF/SARPO, 29 p.

Osborn, F.V. 2002. *J. Wildl. Manage.*, **66**: 674-677.

Owen, L.N. 1955. *Vet. Rec.*, **67**: 561-566.

Owens, N.W. 1977. *Wildfowl*, **28**: 5-14.

Padgett, S.L., J.E. Stokes, R.L. Tucker and L.G. Wheaton. 1998. *J. American Animal Hospital Assoc.*, **34**: 437-439.

Palmer, T.K. 1976. *Proc. Vetebr. Pest Conf.*, **7**: 17-21.

Palomares, F., P. Gaona, P. Ferreras and M. Delibes. 1995. *Conserv. Biol.*, **9**: 295-304.

Parfitt, D.E and G.J. Fox. 1986. *Can. J. Plant Sci.*, **66**: 19-23.

Parfitt, D.E. 1984. *Can. J. Plant Sci.*, **64**: 37-42.

Parfitt, D.E., G.J. Fox and J.D. Brosz. 1986. *Can. J. Plant Sci.*, **66**: 11-17.

Parkes, J.P. 1990. *J. Royal Soc. New Zealand*, **20**: 297-304.

Parkes, J.P. 1996. *Wildl. Biol.*, **2**: 179-184.

Parkes, J.P., G. Nugent and B. Warburton. 1996. *Wildl Biol.*, **2**: 171-177.

Parkhurst, J. A., R.P. Brooks and D.E. Arnold. 1987. *Wildl. Soc. Bull.*, **15**: 386-394.

Parrott, D and G. Watola. 2008. *Crop Protec.*, **27**: 632-637

Pearson, E.W. 1967. *Proc. Vertebr. Pest Conf.*, **3**: 79-86.

Pearson, E.W., P.R. Skon and G.W. Corner. 1967. *J. Wildl. Manage.*, **31**: 502-506.

Pearson, R.G. 1972. The avian brain. Academic Press, London, 658 p.

Pech, R.P and G.M. Hood. 1998. *J. Appl. Ecol.*, **35**: 434-453.

Pech, R.P., A.R.E. Sinclair and A.E. Newsome. 1995. *Wildl. Res.*, **22**: 55-64.

Peek, J.M. 1966. *J. Am. Vet. Med. Assoc.*, **149**: 950-952.

Pereira, F.M. 1990. *Acta Hortic.*, **275**: 103-109.

Perumal, R.S and T.R. Subramanian. 1973. *Madras Agric. J.*, **60**: 256-258.

Pfeifer, R.P. 1956. *Agronomy J.*, **48**: 139-141.

Pfeifer, R.P. 1957. *Agronomy J.*, **49**: 338.

Pfeifer, W.K and M.W. Goos. 1982. *Proc. Vertebr. Pest Conf.*, **10**: 54-61.

Phillips, P.C. 1961. The fur trade. Univ. Oklahoma Press, Norman. Vol. 1, 686 p.

Pierce, R.A. 1972. Methods useful in reducing blackbird damage to rice fields. Univ. Arkansas Coop. Ext. Service and U.S. Dept. Agriculture, **496**, 4 p.

Piperno, E. 1965. Effects of various paralyzers, tranquilizers, and anesthetics on white-tailed deer. Midwest Assn. of Game and Fish Comm. Conf., Lansing, Mich., 6 p.

Polson, J.E. 1983. Application of aversion techniques for the reduction of losses to beehives by black bears in northeastern Saskatchewan, SRC Publication, Ottawa, Ontario, Canada, 25 p.

Poole, D.W and I.G. McKillop. 2002. *Mamm. Rev.*, **32**: 51-57.

Poole, J.H. 1999. *Animal Behav.*, **58**: 185-193.

Popenoe, P.B. 1913. Date growing in the old world and the new. West India Gardens, Altadena, CA, 316 p.

Popilskis, S.J., M.C. Oz, P. Gorman, A. Florestal and D.F. Kohn. 1991. *Lab. Anim. Sci.*, **41**: 51-53.

Porter, W.F. 1983. *Wildl. Soc. Bull.*, **11**: 325-327.

Post, K., L.E. Evans and W. Jochle. 1988. *Theriogenol.*, **29**: 1233-1243.

Potocnik, H., F. Kljun, J. Racnik, T. Skrbinsek, M. Adamic and I. Kos. 2002. *Acta Theriologica*, **47**: 211-219.

Potvin, N and J.M. Bergeron. 1981. *Phyto Protec.*, **62**: 22-23.

Powell, R.A and G. Proulx. 2003. *ILAR Journal*, **44**: 259-276.

Powell, T and B. Powell. 1977. Your Garden Homestead. Houghton Mifflin Co., Boston, MA, 272 p.

Powlesland, R.G. 1998. Impact of aerial 1080 operations on robins and tomtits at Pureora. *Department of Conserv. Newsletter Series: Rare Bits*, **31**: 1-2.

Powlesland, R.G., J.J.W. Knegtmans and I.S.J. Marshall. 1999. *New Zealand J. Ecol.*, **23**: 149-159.

Preusser, S.E., T.W. Seamans, A.L. Gosser and R.B. Chipman. 2008. *Proc. Vertebr. Pest Conf.*, **28**: 66-73.

Priddel, D., N. Carlile and R. Wheeler. 2000. *Biol. Conserv.*, **94**: 115-125.

Prior, R. 1983. Trees and deer: How to cope with deer in forest field and garden. Batsford, London, 208 p.

Pritts, M.P. 2001. *New York Fruit Quart.*, **9**: 5-7.

Proctor, D.L. 1994. Assessed through http://www.fao.org/docrep/t1838e/t1838e00.htm.

Proulx, G. 1999. **In**: Mammal trapping. Alpha Wildlife Research and Management Ltd., Alberta, Canada, 46 p.

Provenza, F.D. 1996. *J. Anim. Sci.*, **74**: 2010-2020.

Provenza, F.D., E.A. Burrit, T.P. Clausen, J.P. Bryant, P.B. Reichardt, R.A. Distel. 1990. *Am. Nat.*, **136**: 810-828.

Purdy, K.G., W.F. Siemer, G.A. Pomerantz and T.L. Brown. 1987. *Proc. Eastern Wildl. Damage Cont. Conf.*, **3**: 118-127.

Qian, S and Z. Wang. 1984. *Ann. Rev. Pharmacol. Toxicol.*, **24**: 329-360.

Quigley, T.M., H.R. Sanderson, A.R. Tiedemann and M.I. McInnes. 1990. *Rangelands*, **12**: 152-155.

Quy, R.J., D.S. Shepherd and I.R. Inglis. 1992. *Crop Protec.*, **11**: 14-20.

Radwan, M.A and G.L. Crouch. 1978. *J. Chem. Ecol.*, **4**: 675-683.

Radwan, M.A., G.L. Crouch, C.A. Harrington and W.D. Ellis. 1982. *J. Chem. Ecol.*, **8**: 241-253.

Rai, A.K., P.K. Khare and P. Mor. 2009. *Contemporary Engineering Sci.*, 2: 265-268.

Rajasekaran, M., J.S. Bapna, S. Lakshmanan, A.G.R. Nair, A.J. Veliath and M. Panchanadam. 1988. *J. Ethnopharmacol.*, 24: 115-121.

Rappole, J.H., A.R. Tipton, A.H. Kane and R.H. Flores. 1989. *Proc. Great Plains Wildl. Damage Control Workshop*, 9: 129-132.

Rasmussen, L.E.L. 1994. In: Medical management of the elephant. Indira Publishing House, West Bloomfield, Michigan, USA, pp. 207-217.

Rasmussen, L.E.L., J. Anthony, H. Martin and D.L. Hess. 1996. *J. Mamm.*, 77: 422-439.

Rathore, A. 1989. Feral Pig Management Programme. Part 2, Blue Mountains District, Oberon. Internal report, N.S.W. National Parks and Wildlife Service. Cited in *Wildl. Res.*, 1993, 20: 15-22.

Redfern, R and J.E. Gill. 1980. *J. Hygiene*, 84: 263-268.

Reeser, D.W. 1993. *Proc. Conf. Res. Resource Manage. in Parks and on Public Lands, George Wright Soc.*, 7: 431-436.

Reidinger Jr., R.F and J.L. Libay. 1979. *Proc. Bird Control Sem.*, 201-206.

Reidinger, R.F and J.R. Mason. 1983. In: Vertebrate pest control and management materials. Am. Soc. Testing and Measurement, Philadelphia, PA, pp. 20-42.

Reidy, M.M., T.A. Campbell and D.G. Hewitt. 2008. *J. Wildl. Manage.*, 72: 1012-1018.

Rendall, D. 2003. *J. Acoust Soc. Am.*, 113: 3390-3402.

Renfrew, R.B and A.M. Saavedra. 2007. *Ornitologia Neotropical*, 18: 61-73.

Ribot, R.F.H., M.L. Berg, K.L. Buchanan and A.T.D. Bennett. 2011. *EMU Aust. Ornithol.*, 111: 360-367.

Riede, T and K. Zuberbuhler. 2003. *J. Acoust. Soc. Am.*, 114: 1132-1142.

Ritter, C.M. 1978. *Proc. Eastern Pine and Meadow Vole Symp.*, 2: 10p.

Roach, F.A. 1985. Cultivated Fruits of Britain. Basil Blackwell, 349 p.

Robinson, W.B and V.T. Harris. 1960. *Amer. Cattle Producer*, 42: 2.

Rogers, J.G. 1978. In: Flavour chemistry of animal foods. American Chemical Society Symposium Series No. 67, American Chemical Society, Washington, D.C., pp. 150-165.

Rogers Jr. J.G. 1974. *J. Wildl. Manage.*, 38: 418-423.

Rogler, J.C., H.R.R. Ganduglia and R.G. Elkin. 1985. *Nutr. Res.*, 5: 1143-1151.

Romin, L.A and J.A. Bissonette. 1996. *Wildl. Soc. Bull.*, 24: 276-283.

Rooke, I.J. 1984. *J. Agric.* (Australia), 25: 19-21.

Rose, R.K., N.A. Slade and J.H. Honacki. 1977. *Acta Theriologica*, 22: 296-307.

Rosenberry, C.S., L.I. Muller and M.C. Conner. 2001. *Wildl. Soc. Bull.*, 29: 754-757.

Rosenthal, G.A and D.H. Janzen. 1979. Herbivores: Their interaction with plant secondary metabolites. Academic Press, New York, 718 p.

Ross, J and A.M. Tittensor. 1986. *Phil. Trans. R. Soc. Lond. B*, **314**: 599-606.

Ross, J., S. Hix, G. Guilford, S. Thompson, L. Shapiro, D. MacMorran and C. Eason. 2011. *New Zealand J. Zool.*, **38**: 185-188.

Rossler S.T., T.M. Gehring, R.N. Schultz, M.T. Rossler, A.P. Wydeven and J.E. Hawley. 2012. *Wildl. Soc. Bull.*, **36**: 176-184.

Roughton, R.D. 1975. *J. Am. Vet. Med. Assoc.*, **167**: 574-576.

Rousi, M., J. Tahvanainen, R. Julkunen-Tutto and U. Kurten. 1988. *Proc. Vertebr. Pest Conf.*, **13**: 180-182.

Roussel, Y.E and R. Patenaude. 1975. *J. Wildl. Manage.*, **39**: 634-636.

Rowe, F.P., A. Bradfield and T. Swinney. 1985. *J. Hygiene*, **95**: 623-627.

Rowe, F.P., C.J. Plant and A. Bradfield. 1981. *J. Hygiene*, **87**: 171-177.

Rowe, J.J. 1971. *Q. J. For.*, **65**: 148-157.

Roy, J and J.M. Bergeron. 1990. *J. Chem. Ecol.*, **16**: 801-808.

Royall, W.C and J.A. Neff. 1961. *Trans. North Am. Wildl. Nat. Resour. Conf.*, **26**: 234-238.

Rozin, P., L. Gruss and G. Berk. 1979. *J. Comp. Physiol. Psychol.*, **93**: 1001-1014.

Rushton, S.P., G.W. Barreto, *et al.* 2000. *J. Appl. Ecol.*, **37**: 475-490.

Sahr, D.P and F.F. Knowlton. 2000. *J. Wildl. Soc. Bull.*, **28**: 597-605.

Saini, M.S and V.R. Parshad. 1992. *Int. Biodet. Biodeg.*, **30**: 87-96.

Salmon, T.P and F.S. Conte. 1981. Control of bird damage at aquaculture facilities. Univ. California Coop. Ext. Service and U. S. Fish and Wildlife Service Wildlife Management, **475**, 11 p.

Salmon, T.P and W.P. Gorenzel. 2005. Cliff swallows. Publication 7482. University of California Agriculture and Natural Resources, Davis, California, USA.

Salmon, T.P., F.S. Conte and W.P. Gorenzel. 1986. Bird damage at aquaculture facilities. Inst. Agric. Nat. Resour., Univ. Nebraska, Lincoln, USA, 9 p.

Salmon, T.P., W.P. Gorenzel and A.B. Pearson. 2000. An operational crow control program using broadcast calls. Final report to California Department of Food and Agriculture. Vertebrate Pest Control Research Advisory Council. Sacramento, CA. Contract No. 96-0486-III.

Salmon, T.P., W.P. Gorenzel, A.B. Pearson and S.R. Ryan. 1999. A test of broadcast calls to reduce crow damage in almonds. Final report to California Department of Food and Agriculture. Vertebrate Pest Control Research Advisory Council. Sacramento, CA. Contract No. 96-0486-II.

Saul, E.K. 1967. *J. Royal Aeronautical Soc.*, **71**: 366-375.

Saunders, G., K. Barry, P. Parker. 1990. *Aust. Wildl. Res.*, **17**: 525-533.

Savarie, P.J. 1976. *Proc. Vertebr. Pest Conf.*, **7**: 178-184.

Savarie, P.J., D.J. Hayes, R.T. McBride and J.D. Roberts. 1979. **In**: Avian and mammals toxicology. ASTM STP **693**, pp. 69-79.

Savarie, P.J., D.S. Vice, L. Bangerter, K. Dustin, P.J. William, P.M. Thomas and S.F. Blom. 2004. *Proc. Vertebr. Pest Conf.*, **21**: 64-69.

Scaife, M. 1976. *Anim. Behav.*, **24**: 200-204.

Schafer Jr., E.W. 1978. *Proc. Vertebr. Pest Conf.*, **8**: 32-35.

Schafer, E.W and L.L. Marking. 1975. *J. Wildl. Manage.*, **39**: 807-811.

Schafer, E.W., R.B. Brunton and D.J. Cunningham. 1973. *Toxicol. Appl. Pharmacol.*, **26**: 532-538.

Schafer, E.W., W.A. Bowles Jr. and J. Hurlbut. 1983. *Arch. Environ. Contam. Toxicol.*, **12**: 355-382.

Schafer Jr., E.W and R.B. Brunton. 1971. *J. Wildl. Manage.*, **35**: 569-572.

Schehka S., K.H. Esser and E. Zimmerman. 2007. *J. Comp. Physiol. A*, **193**: 845-852.

Schemnitz, S. 1994. **In**: Prevention and control of wildlife damage. Cooperative Extension, University of Nebraska, Lincoln, NE, pp. 81-83.

Schlageter, A and D. Haag-Wackernagel. 2012. *J. Agric. Sci.*, **4**: 61-68.

Schley, L., M. Dufrene, A. Krier and A.C. Frantz. 2008. *Europ. J. Wildl. Res.*, **54**: 589-599.

Schultz, B.O. 1986. *Indian For.*, **12**: 891-899.

Schultz, R.N., K.W. Jonas, L.H. Skuldt and A.P. Wydeven. 2005. *Wildl. Soc. Bull.*, **33**: 142-148.

Schwab, R.G. 1978. *Proc. Vertebr. Pest Conf.*, **8**: 255-259.

Schwan, T.G. 1986. *Afr. J. Ecol.*, **24**: 31-35.

Scott, J.D and T.W. Townsend. 1985. *Wildl. Soc. Bull.*, **13**: 234-240.

Scott, M.E. 1987. *Parasitol.*, **95**: 111-124.

Scott, M.E. 1990. *Parasitol.*, **101**: 75-92.

Sealander, J.A and D. James. 1958. *J. Mamm.*, **39**: 215-223.

Seamans, T.W and G.E. Bernhardt. 2004. *Proc. Vertebr. Pest Conf.*, **21**: 104-106.

Seamans, T.W and J.L. Belant. 1999. *Wildl. Soc. Bull.*, **27**: 729-733.

Seamans, T.W. 2004. *Ohio J. Sci.*, **104**: 136-138.

Seamans, T.W., B.F. Blackwell and J.D. Cepek. 2002. *Int. J. Pest Manage.*, **48**: 301-306.

Seamans, T.W., C.D. Lovell, R.A. Dolbeer and J.D. Cepek. 2001. *Wildl. Soc. Bull.*, **29**: 1061-1066.

Seamans, T.W., C.R. Hicks and K.J. Preusser. 2007a. Assessed through http:// digitalcommons.unl.edu/birdstrike2007/15

Seamans, T.W., S.C. Barras and G.E. Bernhardt. 2007b. *Int. J. Pest Manage.*, **53**: 45-51.

Seamans, T.W., S.C. Barras and J.Z. Patton. 2003. Assessed through http:// digitalcommons.unl.edu/birdstrike2003/12.

Seamans, T.W., S.W. Young and J.D. Cepek. 2000. Response of roosting turkey vultures to a hanging vulture effigy. Federal Aviation Administration Interim Report DTFA03-99-X-90001, Task 3, Experiment 5, Atlantic City, NJ.

Seamark, R.F. 2001. *Reprod. Fert. Dev.*, **13**: 705-711.

Seidensticker, J. 1984. Assessed through http://pdf.usaid.gov/pdf_docs/ PNAAQ728.pdf.

Seiler, G.J and C.E. Rogers. 1987. *Agric. Ecosyst. Environ.*, **20**: 59-70.

Seubert, J.L. 1964. *Proc. Vertebr. Pest Conf.*, **2**: 150-159.

Seubert, J.L. 1965. Biological studies of the problems of bird hazard to aircraft. U.S. Dep. Inter., Bur. Sport Fish and Wildl., Div. of Wildl. Res., Washington, D.C., 27 p.

Severinghous, C.W. 1950. *Cornell Vet.*, **49**: 276-281.

Seyfarth, R.M and D.L. Cheney. 2003. *Ann. Rev. Psychol.*, **54**: 145-173.

Seyfarth, R.M., D.L. Cheney and P. Marler. 1980. *Anim. Behav.*, **28**: 1070-1094.

Shagli, R., A. Matityahu, S.J. Gaunt and R. Jones. 1990. *Mol. Reprod. Dev.*, **25**: 286-296.

Shah, P.S., L. Clark and J.R. Mason. 1991. *Pestic. Biochem. Physiol.*, **40**: 169-175.

Shalter, M.D. 1978. *Anim. Behav.*, **26**: 1219-1221.

Shapiro, L., C.T. Eason, E. Murphy, P. Dilks, S. Hix, S.C. Ogilvie and D. MacMorran. 2010. *Proc. Vertebr. Pest Conf.*, **24**: 108-114.

Sheafor, S.E and C.G. Couto. 1999. *J. American Animal Hospital Assoc.*, **35**: 38-46.

Sheil, D., E.P. Francis and J.Z. Roderick. 2010. Biodiversity conservation in certified forests. European Tropical Forest Research Network. Issue No. 51. Tropenbos International, Wageningen, The Netherlands, 204 p.

Shellam, G.R. 1994. *Reprod. Fert. Dev.*, **6**: 401-409.

Shennan, J.P. 1960. *New Zealand J. Agric.*, **101**: 145.

Shimizu, Y., A. Inagaki, Y. Taneda, M. Takematsu, Y. Ohtake and M. Natamori. 1988. *Bull. Appl. Ornithol.*, **8**: 21-48.

Shirota, Y., M. Sanada and S. Masaki. 1983. *Appl. Ent. Zool.*, **18**: 545-549.

Shivik, J.A and D.J. Martin. 2000. Assessed through http://digitalcommons.unl. edu/icwdm_wdmconfproc/20

Shivik, J.A. 2004. *Sheep and Goat Res. J.*, **19**: 64-71.

Shivik, J.A., A. Treves and P. Callahan. 2003. *Conserv. Biol.*, 17: 1531-1537.

Shore, R.F., J.D.S. Birks and P. Freestone. 1999. *New Zealand J. Ecol.*, **23**: 199-206.

Shorey, H.H., C.B. Sisk and R.G. Gerber. 1996. *Environ. Entomol.*, **25**: 446-451.

Shubert, G.H and R.S. Adams. 1971. Reforestation practices for conifers in California. Calif. Div. Forestry, Sacramento, 359 p.

Shumake, S.A., A.L. Kolz, R.F. Reidinger and M.W. Fall. 1979. In: Vertebrate pest control and management materials. ASTM STP 680, Philadelphia, pp. 29-38.

Simon, G. 2008. *Int. J. Horti. Sci.*, **14**: 107-111.

Sinclair, A.R.E. 2003. *Phil. Trans. R. Soc. Lond. B*, **358**: 1729-1740

Sinclair, A.R.E., R.P. Pech, C.R. Dickman, D. Hik, P. Mahon and A.E. Newsome. 1998. *Conserv. Biol.*, **12**: 564-575.

Singhar, S.A. 2002. *Indian Forester*, **128**: 1113-1118.

Singla, N and V.R. Parshad. 2009. *Int. Pest Control*, **51**: 36-39.

Singleton, G.R and D.M. Spratt. 1986. *Aust. J. Zool.*, **34**: 677-681.

Singleton, G.R. 1994. *Proc. Vertebr. Pest Conf.*, **16**: 301-307.

Singleton, G.R., L.E. Twigg, K.E. Weaver and B.J. Kay. 1991. *Wildl. Res.*, **18**: 275-283.

Singleton, G.R., Sudarmaji, J. Jacob, C.J. Krebs. 2005. *Agric. Ecosyst. Environ.*, **107**: 75-82.

Sitati, N.W and M.J. Walpole. 2006. *Oryx*, **40**: 279-286.

Sitati, N.W., M.J. Walpole and N. Leader-Williams. 2005. *J. Appl. Ecol.*, **42**: 1175-1182.

Slade, N.A., M.A. Eifler, N.M. Gruenhagen and A.L. Davelos. 1993. *J. Mamm.*, **74**: 156-161.

Slater, P.J.B. 1980. In: Bird problems in agriculture. BCPC Publ. Croydon, England, pp. 105-114.

Smith, M., L. Staples, B. Dyer and W. Hunt. 2002. *Proc. Conf. Australian Soc. Sugarcane Technologists*, Cairns, Queensland, Australia, 29 April - 2 May 2002, pp. 228-234.

Smith, M.E. 1984. Repellents and deterrents for black and grizzly bears. Progress report. University of Montana, USA. Cited in *Pachyderm*, **20**: 55-64.

Smith, P., I.R. Inglis, D. Cowan, D.R. Kerins and D.S. Bull. 1994. *J. Comp. Psychol.*, **108**: 282-290.

Smuts, G.L. 1973. *Koedoe - African Protected Area Conserv. Sci..*, **16**: 175-180.

Solman, V.E.F. 1976. *Proc. Bowling Green Bird Control Sem.*, Bowling Green State Univ., **17**: 83-88.

Spalding, W.A. 1885. The orange: Its culture in California. Riverside: Press and Horticulturalist Steam Print, 97 p.

Spelman, L.H., P.W. Sumner, W.B. Karesh and M.K. Stoskopf. 1997. *J. Zool. Wildl. Med.*, **28**: 418-423.

Spillstoesser, W.E. 1984. Vegetable Growing Handbook. AVI Publishing Co. Inc., Westport, CT, 325 p.

Spratt, D.M. 1990. Int. *J. Parasitol.*, **20**: 543-550.

Sprock, W.L., W.E. Howard and F.C. Jacob. 1967. *J. Wildl. Manage.*, **31**: 729-741.

Spurr, E.B and P.G. McGregor. 2003. Potential invertebrate antifeedants for toxic baits used for vertebrate pest control: A literature review. Science for conservation 232, New Zealand, 32 p.

Spurr, E.B and R.E.R. Porter. 1998. *Proc. Aus. Vertebr. Pest Conf.*, **11**: 295-299.

Stephen, W.J.D. 1961. *North Amer. Wildl. Conf.*, **26**: 98-111.

Stephenson, B.H., E.O. Minot and D.P. Armstrong. 1999. *New Zealand J. Ecol.*, **23**: 233-240.

Sterner, R.T and R.E. Mauldin. 1995. *Arch. Environ. Cont. Toxicol.*, **28**: 519-523.

Stevens, G.R and L. Clark. 1998. *Int. Biodet. Biodeg.*, **42**: 153-160.

Stevens, G.R., J. Rogue, R. Weber and L. Clark. 2000. *Int. Biodet. Biodeg.*, **45**: 129-137.

Stevens, G.R., L. Clark and R.A. Weber. 1998. *Proc. Vertebr. Pest Conf.*, **18**: 74-76.

Stevens, V.C. 1986. *Immunol. Today*, 7: 369-374.

Stickley, A.R., R.T. Mitchell, R.G. Heath, C.R. Ingram and E.L. Bradley. 1972. *J. Wildl. Manage.*, **36**: 1313-1316.

Stickley Jr. A.R and C.R. Ingram. 1973. *Proc. Bird Control Sem.*, **6**: 41-46.

Stickley Jr. A.R. and O. King Jr. 1993. *Eastern Wildl. Damage Control Conf.*, **6**: 89-92.

Stirling, I. 1968. *Can. J. Zool.*, **46**: 405-408.

Stoddart, D.M. 1980. The ecology of vertebrate olfaction. Chapman and Hall, New York.

Stoeckeler, J.H and H.F. Scholz. 1956. *J. Forestry*, **54**: 183-184.

Stone, C.P., W.F. Shake and D.J. Langowski. 1974. *Wildl. Soc. Bull.*, **2**: 135-139.

Stone, R.J. 1979. *Proc. Bird Control Sem.*, **8**: 90-95.

Stone, W.B., J.C. Okoniewski and J.R. Stedelin. 1999. *J. Wildl. Diseases*, **35**: 187-193.

Stoner, T.I., G.H. Vansell and B.D. Moses. 1938. *J. Wildl. Manage.*, **2**: 171-178.

Stout, J.F and E.R. Schwab. 1979. *Proc. Bird Control Sem.*, **8**: 96-110.

Stout, J.F., C.R. Wilcox and L.E. Creitz. 1969. *Behaviour*, **34**: 29-41.

Stucky, J.T. 1973. *Proc. Bird Control Sem.*, **6**: 195-197.

Sukumar, R. 1989. The Asian elephant: Ecology and management. Cambridge University Press, Cambridge, 251 p.

Sullivan, T.P and D.R. Crump. 1984. *J. Chem. Ecol.*, **10**: 1809-1821.

Sullivan, T.P and D.R. Crump. 1986. *J. Chem. Ecol.*, **12**: 729-739.

Sullivan, T.P. 1978. *Proc. Vertebr. Pest Conf.*, **8**: 237-250.

Sullivan, T.P. 1986. *J. Mammal.*, **67**: 385-388.

Sullivan, T.P., D.R. Crump and D.S. Sullivan. 1988b. *J. Chem. Ecol.*, **14**: 379-389.

Sullivan, T.P., D.R. Crump and S.S. Sullivan. 1988a. *J. Chem. Ecol.*, **14**: 363-378.

Sullivan, T.P., L.O. Nordstrom and D.S. Sullivan. 1985a. *J. Chem. Ecol.*, **11**: 903-919.

Sullivan, T.P., L.O. Nordstrom and D.S. Sullivan. 1985b. *J. Chem. Ecol.*, **11**: 921-935.

Summers, R.W and G. Hillman. 1990. *Crop Protec.*, **9**: 459-462.

Summers, R.W. 1985. *Crop Protec.*, **4**: 520-528.

Sushil, S.N., K.P. Singh, J. Stanley, J.C. Bhatt and H.S. Gupta. 2009b. VL-WAR a wild animal repellent. VPKAS publication No. 56/2009.

Sushil, S.N., K.P. Singh, J. Stanley, J.C. Bhatt and H.S. Gupta. 2009a. *ICAR News*, **15**: 22.

Swamy, S.T.P.V.J., N. Shivanarayan and M.H. Ali. 1980. *J. Bombay. Nat. Hist. Soc.*, **77**: 335-336.

Swift, B.L. 2000. *Proc. Wildl. Damage Manage. Conf.*, **9**: 307-321.

Swihart, R.K. 1991. *Ecol. Appl.*, **1**: 98-103.

Swihart, R.K., J.J. Pignatello and M.J. Maiitina. 1991. *J. Chem. Ecol.*, **17**: 767-777.

Szolcsanyi, J., H. Sann and F.K. Pierau. 1986. *Pain*, **27**: 247-260.

Tahvanainen, J., E. Helle, R. Julkunen-Tutto and A. Lavola. 1985. *Oecologia*, **65**: 319-323.

Tattersall, F and S. Whitbread. 1994. *J. Zool.* (London), **233**: 309-314.

Taylor, D and L. Katahira. 1988. *Wildl. Soc. Bull.*, **16**: 297–299.

Taylor, R.D and R.B. Martin. 1987. *Environ. Manage.*, **11**: 327-334.

Taylor, R.D. 1982. *Zimbabwe Agric. J.*, **79**: 179-184.

Taylor, R.D. 1999. A review of problem elephant policies and management options in southern Africa. Human-Elephant Conflict Task Force, International Union for the Conservation of Nature and Natural Resources, Nairobi, Kenya.

Tchamba, M.N. 1996. *Biol. Conserv.*, **75**: 35-41.

Telfer, S., A. Holt, *et al.* 2001. *Oikos*, **95**: 31-42.

Terborgh, J., Lambert, *et al.* 2001. *Science*, **294**: 1923-1926.

Ternent, M.A and D.L. Garshelis. 1999. *Wildl. Soc. Bull.*, **27**: 720-728.

The Wildlife Protection Act, 1972. 2005. Bare act with short notes. Universal law publishing pvt ltd., New Delhi, 153p.

Theissen, G.J., E.A.G. Shaw, R.D. Harris, J.B. Gollop and H.R. Webster. 1957. *J. Acoustical Soc. Amer.*, **29**: 1301-1306.

Theuerkauf, J., S. Rouys, H. Jourdan and R. Gula. 2011. *Ann. Zool. Fennici.*, **48**: 308-318.

Thijssen, H.H.W. 1995. *Pestic. Sci.*, **43**: 73-78.

Thomas, W.B. 1996. *J. Nematol.*, **28**: 586-589.

Thomas, A.D and F.F. Kolbe. 1942. *J. South Afr. Vet. Med. Assoc.*, **13**: 1-11.

Thomas, M and P. Ross. 2007. Breakdown of cyanide and cholecalciferol in Feratox® and Feracol® possum baits. DOC Research and Development Series 288. Wellington, Department of Conservation, 28 p.

Thomas, M.D., F.W. Maddigan, J.A. Brown and M. Trotter. 2003. *New Zealand Plant Protec.*, **56**: 77-80.

Thompson, H.V. 1958. Rabbit control in Australia and New Zealand. *New Zealand Agric.*, **65**: 388-392 & 440-444.

Thompson, I.D and A.L. Macauley. 1987. *Can. Field Nat.*, **101**: 608-610.

Thouless, C.R and J. Sakwa. 1995. *Biol. Conserv.*, **72**: 99-107.

Thouless, C.R. 1994. Conflict between humans and elephants in Sri Lanka. Report for the GEF, EDG, Oxford, UK, 33p.

Thumen, A., P. Albrecht, C. Dodt and A. Moser. 2002. *Aktuelle Neurologie*, **29**: 409-413.

Tiedemann, A.R., T.M. Quigley, L.D. White and W.S. Lauritzen, J.W. Thomas and M.I. McInnes. 1999. Assessed through http://www.fs.fed.us/pnw/pubs/rp_510.pdf

Tierson, W.C. 1969. *J. Wildl. Manage.*, **33**: 922-926.

Tigner, J.R and W.A. Bowles Jr. 1964. *J. Wildl. Manage.*, **28**: 748-751.

Till, J.A and F.F. Knowlton. 1983. *J. Wildl. Manage.*, **47**: 1018-1025.

Tillman, E.A., A.C. Genchi, J.R. Lindsay, J.R. Newman and M.L. Avery. 2004. *Proc. Vertebr. Pest Conf.*, **21**: 126-129.

Tillman, E.A., J.S. Humphrey and M.L. Avery. 2002. *Proc. Vertebr. Pest Conf.*, **20**: 123-128.

Tim, R.M. 1983. **In**: Prevention and control of wildlife damage. Univ. Nebraska Press, Lincoln, NE, pp. 27-28.

Timm, R.M. 1994a. **In**: Prevention and control of wildlife damage. Lincoln, NE, USA: University of Nebraska Co-op. Ext. Serv., pp. 23-61.

Timm, R.M. 1994b. Assessed through http://www.icwdm.org/handbook/allPDF/ACTIVE.PDF

Tipton, A.R., J.H. Rappole, *et al.*, 1989. *Proc. Great Plains Wildl. Damage Control Workshop*, **9**: 126-128.

Tipton, K.W., E.H. Floyd, J.G. Marshal and J.B. Mcdevitt. 1970. *Agron. J.*, **62**: 211-213.

Tobin, M.E and R.W. DeHaven. 1984. *Agric. Ecosyst. Environ.*, **11**: 291-297

Tobin, M.E. 1998. *Proc. Vertebr. Pest Conf.*, **18**: 67-70.

Tobin, M.E., P.P. Woronecki, R.A. Dolbeer and R.L. Bruggers. 1988. *Wildl. Soc. Bull.*, **16**: 300-303.

Tobin, M.E., R.A. Dolbeer and C.M. Webster. 1989. *Crop Protec.*, **8**: 461-465.

Todd, A.W. 1987. *Wildl. Soc. Bull.*, **15**: 372-380.

Tomlin, C. 1994. The Pesticide manual, incorporating the agrochemicals handbook. A world compendium. 10th Edition. British Crop Protection Council, Farnham, U.K., 1341 p.

Travaini, A., R.M. Peck and S.C. Zapata. 2001. *Wildl. Soc. Bull.*, **29**: 1089-1096.

True, G.H. 1932. *Calif. Fish and Game*, **18**:156-165.

Tsachalidis, E., C. Sokos, P. Birtsas and N. Patsikas. 2006. *Proc. Naxos Intl Conf. on Sustainable Management and Development of Mountainous and Island Areas.* Vol. II., pp. 325–329.

Tucker, R.E., W. Majak, P.D. Parkinson and A. McLean. 1976. *J. Range Manage.*, **29**: 486-489.

Tukey, L.D. 1959. *Proc. American Soc. Horti. Sci.*, **74**: 30-39.

Tundale-Biscoe, C.H. 1991. *Reprod. Fert. Dev.*, **3**: 339-343.

Tuyttens, F.A.M and D.W. MacDonald. 1998. *Anim. Welf.*, **7**: 339-364.

Twedt, D.J. 1980. *Environ. Conserv.*, **7**: 217-221.

Twigg, L.E., G.R. Singleton and B.J. Kay. 1991. *Wildl. Res.*, **18**: 265-274.

Tyas, J.A., P.J. Hofman, S.J.R. Underhill and K.L. Bell. 1998. *Scientia Hortic.*, **72**: 203-213.

Uhler, F.M and S. Creech. 1939. Protecting field crops from waterfowl damage by-means of reflectors and revolving beacons. U.S. Bureau of Biological Survey Wildlife Publicaiton. **149**. Washington, DC., 5 p.

Uthamasamy, S., K.M. Balasubramanian and N.M. Ramaswamy. 1982. *Int. Rice Newsl.*, **7**: 16.

Vandenbergh, J.G and D.E. Davis. 1962. *J. Wildl. Manage.*, **26**: 366-371.

vanDyk, B.W and D. Saayman. 1989. *Deciduous Fruit Grower*, **39**: 276-219.

vanRensburg, P.J.J., J.D. Skinner and R.J. vanAarde. 1987. *J. Appl. Ecol.*, **24**: 63-73.

Vaudry, A.L. 1979. Bird control for agricultural lands in British Columbia. Publications--British Columbia Ministry of Agric., 19 p.

VerCauteran, K.C and S.E. Haganstorm. 1998. *J. Wildl. Manage.*, **62**: 280-285.

VerCauteran, K.C and S.E. Haganstorm. 2002. *Proc. Natl. Bowhunting Conf.*, **1**: 51-55.

VerCauteran, K.C., M.J. Lavelle and S. Hygnstrom. 2006. *Wildl. Soc. Bull.*, **34**: 191-200.

VerCauteren, K., N. Seward, D. Hirchert, M. Jones, and S. Beckerman. 2005a. *Wildl. Damage Manage. Conf.*, **11**: 286-293.

VerCauteren, K.C., J.A. Shivik, M.J. Lavelle. 2005b. *Wildl. Soc. Bull.*, **33**: 1282-1287.

VerCauteren, K.C., J.M. Gilsdorf, S.E. Hygnstrom, P.D. Fioranelli, J.A. Wilson and

S. Barras. 2006. *J. Wildl. Manage.*, **34**: 371-374.

Vernet-Maury, E. 1980. In: Olfaction and Taste. IRL Press, Washington, D.C., 407 p.

Vettorazi, G and I. MacDonald. 1988. Sucrose: nutritional and safety aspects. Springer-Verlag, New York, 192 p.

Vickery, J.A and R.W. Summers. 1992. *Crop Protec.*, **11**: 480-484.

Vincent, C and M.J. Lareau. 1993. *Crop Protec.*, **12**: 397-399.

Vogt, P.F. 1997. *Proc. Great Plains Wildl. Damage Workshop*, **13**: 63-66.

Volle, R.L and G.B. Koelle. 1975. In: The pharmacological basis of therapeutics, Macmillan, New York, pp. 565-574.

Vollrath, F and I. Douglas-Hamilton. 2002. *Naturwissenshaften*, **89**: 508-511.

Waage, J.K and D.J. Greathead. 1988. *Phil. Trans. R. Soc. Lond. B*, **318**: 111-128.

Wade, D.A. 1982. *Proc. Verteb. Pest Conf.*, **10**: 24-33.

Wade, D.A. 1983. In: Prevention and control of wildlife damage. Great Plains Agric. Council, Wildl. Resources Committee and Nebraska Coop. Exten. Serv. Univ. of Nebraska, Lincoln, pp. C 31-41.

Wager-Page, S.A and J.R. Mason. 1996. *J. Wildl. Manage.*, **60**: 917-922.

Wager-Page, S.A., G. Epple and J.R. Mason. 1997. *J. Wildl. Manage.*, **61**: 235-241.

Waldien, D.L., M.M. Cooley, J. Weikel, J.P. Hayes, C.C. Maguire, T. Manning and T.J. Maier. 2004. *Wildl. Soc. Bull.*, **32**: 1260-1268.

Waldvogel, J.A. 1989. *Curr. Ornithol.*, **6**: 269-321.

Wallach, J.D. 1966. *J. Am. Vet. Med. Assoc.*, **149**: 871-874.

Walsh, A.M and T.J. Wardlaw. 2005. *Proc. Aust. Vetebr. Pest Conf.*, **13**: 48-55.

Walton, M.T and C.A. Field. 1989. *Eastern Wildl. Damage Control Conf.*, **4**: 87-94.

Wang, D., N. Li, M. Liu, B. Huang, Q. Liu and X. Liu. 2011. *Physiol. Behav.*, **104**: 1024-1030.

Wang, J and F.D. Provenza. 1996. *J. Anim. Sci.*, **74**: 2349-2354.

Warburton, B. 1990. *Aust. Wildl. Res.*, **17**: 541-546.

Ward, J.G. 1978. Tests of the Syncrude bird deterrent device for use on a tailings pond. Rep. from LGL Ltd. Edmonton, Alb., for Syncrude Canada Ltd., Edmonton, Alb., 115 p.

Warder, J.A. 1867. American pomology: Apples. Orange Judd and Co., New York, 744 p.

Watanabe, A., K. Nakumura and S. Matsuoka. 1988. *Jpn J. Appl. Entomol. Zool.*, **32**: 104-110.

Watkins, R.W., E.L. Gill and J.D. Bishop. 1995. *Pestic. Sci.*, **44**: 335-340.

Watts, C.H.S. 2002. In: The mammals of Australia. Sydney: Reed New Holland, pp. 659-660.

Waugh, F.A. 1901. Plums and plum culture. Orange Judd and Co., New York, 371 p.

Waugh, F.A. 1917. The American apple orchard. Orange Judd and Co., New York, 215 p.

Way, R.D. 1968. Breeding for superior cherry cultivars in New York State. ISHS Symposium on Cherries and Cherry Growing, Bonn, Germany, 121-137.

Wright, E.N. 1969. Bird dispersal techniques and their use in Britain. World Conf. Bird Hazards. Nat. Research Council of Canada, pp. 207-213.

Weingartner, D.H and F.C. Cech. 1974. *Proc. West Virginia Acad. Sci.*, **46**: 27-34.

Welch, B.L., E.D. McArthur and J.N. Davis. 1983. *J. Range Manage.*, **36**: 485-487.

Welch, J.F. 1967. *Proc. Vertebr. Pest Conf.*, **3**: 36-40.

Werner, S.J., A. El Hani and J.R. Mason. 1997. *J. Wildl. Res.*, **2**: 146-148.

Werner, S.J., B.A. Kimball and F.D. Provenza. 2008. *Physiol. Behav.*, **93**: 110-117.

Werner, S.J., H.J. Homan, *et al.*, 2005. *Wildl. Soc. Bull.*, **33**: 251-257.

Werner, S.J., J.L. Cummings, S.K. Tupper, J.C. Hurley, R.S. Stahl and T.M. Primus. 2007. *J. Wildl. Manage.*, **71**: 1676-1681.

West, B.C., A.L. Cooper and J.B. Armstrong. 2009. Assessed through www.berrymaninstitute.org/publications

West, S.D. 1985. *J. Mamm.*, **66**: 798-800.

Westall, R.G. 1953. *Biochem. J.*, **55**: 244-248.

Wetherbee, D.K. 1964. Vertebrate pest control by biological means. A.A.A.S. Symp. Pest Control, Montreal, 19 p.

Wetherbee, D.K., R.P. Coppinger, B.C. Wentworth and R.E. Walsh. 1964. *Univ. Massachusetts Agr. Expt. Sta. Bull.*, **543**: 16.

Wiener, J.G and M.H. Smith. 1972. *J. Mamm.*, **53**: 868-873.

Wilcox, J.T., E.T. Ashehoug, C.A. Scott and D.H. VanVuren. 2004. *Trans. Western Sec. Wildl. Soc.*, **40**: 120-126.

Wilkinson, A.E. 1945. Fruits, berries and nuts and how to grow them. The Blakiston Co., Philadelphia, PA, 271 p.

Willan, K. 1986. *South African J. Wildl. Res.*, **16**: 53-57.

Williams, C.N and Y.C. Hsu. 1979. Oil Palm Cultivation in Malaya. University of Malaya Press, Kuala Lumpur, 205 p.

Wilson, C.J and I.G. McKillop. 1986. *Wildl. Soc. Bull.*, **14**: 409-411.

Witmer, G. 2007. **In**: Perspectives in ecological theory and integrated pest management, Cambridge University Press, Cambridge, UK, pp. 393-410.

Witmer, G.W and M.J. Pipas. 1998. *Proc. Vertebr. Pest Conf.*, **18**: 203-207.

Witmer, G.W., F. Boyd and Z. Hillis-Starr. 2007. *Wildl. Res.*, **34**: 108-115.

Wood, B., B.R. Tershy, *et al.* 2002. **In**: Turning the tide: the eradication of invasive species, Invasive Species Specialist Group, IUCN, Gland, Switzerland, pp. 374-380.

Wood, B.J. 1985. *J. Plant Protec. Tropics*, **2**: 67-79.

Woodburne, L.S. 1979. *WinesVines*, **60**: 36.

Woodroffe, G., J. Lawton and W. Davidson. 1990. *Biol. Conserv.*, **51**: 49-62.

Woolf, A and J.L. Roseberry. 1998. *Wildl. Soc. Bull.*, **26**: 515-521.

Woolf, A. 1970. *J. Am. Vet. Med. Assoc.*, **157**: 636-640.

Woronecki, P.P. 1988. *Proc. Vertebr. Pest Conf.*, **13**: 266-272.

Woronecki, P.P., J.L. Guarinoy and J.W. DeGrazioz. 1967. *Proc. Vertebr. Pest Conf.*, **3**: 54-56.

Woronecki, P.P., R.A. Dolbeer, T.W. Seamans and W.R. Lance. 1992. *Proc. Vertebr. Pest Conf.*, **15**: 72-78.

Woulfe, M.R. 1968. Assessed through http://digitalcommons.unl.edu/icwdmbirdcontrol/176

Yih, J.P. 1995. *British Med. J.*, **314**: 276.

York, D.L., J.L. Cummings and K.L. Wedemeyer. 2000. *Northwest. Nat.*, **81**: 11-17.

Young, S.R and G.R. Martin. 1984. *Vision Res.*, **24**: 129-137.

Yui, M. 1988. *Bull. Appl. Ornithol.*, **8**: 13-20.

Zajanc, A. 1962. *Proc. Vertebr. Pest Conf.*, **1**: 190-212.

Zemlicka, D.E and K.J. Bruce. 1991. *Proc. Great plains Wildl. Damage Control Workshop*, **10**: 52-56.

Zemlicka, D.E., D.P. Sahr, P.J. Savarie, F.F. Knowlton, F.S. Blom and J.L. Belant. 1997. *Proc. Great Plains Wildl. Damage Control Workshop*, **13**: 42-45.

Zuberbuhler, K. 2000. *Anim. Behav.*, **59**: 917-927.

Zuberbuhler, K. 2001. *Behav. Ecol. Sociobiol.*, **50**: 414-422.

Zuberbuhler, K., R. Noe, R.M. Seyfarth. 1997. *Anim. Behav.*, **53**: 589-604.

Index